KERNEL SMOOTHING
in MATLAB

Theory and Practice of
Kernel Smoothing

KERNEL SMOOTHING
in MATLAB
Theory and Practice of
Kernel Smoothing

Ivanka Horová

Jan Koláček

Jiří Zelinka

Masaryk University, Czech Republic

World Scientific

NEW JERSEY · LONDON · SINGAPORE · BEIJING · SHANGHAI · HONG KONG · TAIPEI · CHENNAI

Published by

World Scientific Publishing Co. Pte. Ltd.

5 Toh Tuck Link, Singapore 596224

USA office: 27 Warren Street, Suite 401-402, Hackensack, NJ 07601

UK office: 57 Shelton Street, Covent Garden, London WC2H 9HE

British Library Cataloguing-in-Publication Data
A catalogue record for this book is available from the British Library.

ISBN 978-981-4405-48-5

Printed in Singapore.

To our families

Preface

Methods of kernel estimations represent one of the most effective smoothing techniques. These methods are simply to understand and they possess very good statistical properties of the estimators. The book provides a brief comprehensive overview of statistical theory. We are not going very much into details since there exists a number of excellent monographs developing statistical theory (Wand and Jones (1995); Härdle (1990); Müller (1988); Silverman (1986); Simonoff (1996); Scott (1992) *etc.*). Instead of this, the emphasis is given to implementation of presented methods in MATLAB. All created programs are included into a special toolbox which is an integral part of the book. This toolbox contains many MATLAB scripts useful for kernel smoothing of density, distribution function, regression function, hazard function, multivariate density and also for kernel estimation and reliability assessment. The toolbox can be downloaded from the public web page (see Koláček and Zelinka (2012)).

The toolbox is divided into six parts according to chapters of the book. All scripts are included in a user interface and it is easy to manipulate with this interface. Each chapter of the book contains a detailed help for related part of the toolbox, too.

The monograph is intended for newcomers to the field of smoothing techniques and would be also appropriate for a wide audience: advanced graduate and PhD students, researchers from both the statistical science and interface disciplines.

Since 2003 we have taught much of the presented material in graduate courses at Masaryk University, Brno. Our work has been supported by the project Jaroslav Hájek Center for Theoretical and Applied Statistics (MŠMT LC06024, 2006 – 2011).

Drafts of this text have been read by PhD students Dagmar Lajdová, Iveta Selingerová and the graduate student Kateřina Konečná. Their feedback, comments and corrections are very gratefully acknowledged. Our special thanks belong to our colleagues Kamila Vopatová for her useful comments and cooperation and Martin Řezáč for his contribution to the fourth chapter.

Finally, we have to express our sincere thanks to our partners Jaroslav, Veronika and Markéta and our families for their support and encouragement.

Brno, May 2012

Ivanka Horová
Jan Koláček
Jiří Zelinka

Contents

Chapter 1

Introduction

1.1 Kernels and their properties

Kernel smoothing belongs to a general category of techniques for nonparametric curve estimation including nonparametric regression, nonparametric density estimators and nonparametric hazard functions. These estimations depend on a smoothing parameter called a bandwidth which controls the smoothness of the estimate and on a kernel which plays a role of weight function. As far as the kernel function is concerned, a key parameter is its order which is related both to the number of its vanishing moments and to the number of existing derivatives for the underlying curve to be estimated.

In this chapter, we introduce a definition of the kernel and show some of its useful properties. Various aspects of the choice of the kernel function have been discussed, *e.g.*, in Wand and Jones (1995); Müller (1988). As concerns a bandwidth choice – it is a crucial problem in the kernel smoothing and it will be discussed in the next chapters.

Throughout this book the following definition of a kernel is suitable for our considerations.

Definition 1.1. Let ν, k be nonnegative integers, $0 \leq \nu < k$. Let K be a real valued function satisfying $K \in S_{\nu,k}$, where

$$
S_{\nu,k} = \begin{cases} K \in Lip[-1,1], \ \text{support}(K) = [-1,1] \\ \int\limits_{-1}^{1} x^j K(x)dx = \begin{cases} 0, & 0 \leq j < k, j \neq \nu \\ (-1)^\nu \nu!, & j = \nu \\ \beta_k(K) \neq 0, j = k. \end{cases} \end{cases}
\tag{1.1}
$$

Such a function is called a *kernel* of order k. The integral conditions are often called moment conditions. We will use the short notation β_k instead of $\beta_k(K)$ if there cannot be any misunderstanding.

A commonly used kernel function is the Gaussian kernel

$$K(x) = \frac{1}{\sqrt{2\pi}} \exp(-x^2/2).$$

But this kernel has an unbounded support and thus it does not belong to the class $S_{\nu,k}$. For our purpose the kernels with bounded support will be more useful. Figure 1.2 shows the most popular polynomial kernel – the Epanechnikov kernel (see Epanechnikov (1969)).

Example 1.1. Figures 1.1 – 1.4 present some kernels from the class $S_{0,2}$. I_A denotes the indicator function of the set A.

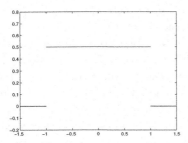

Fig. 1.1 $K(x) = \frac{1}{2}I_{[-1,1]}(x)$, uniform kernel.

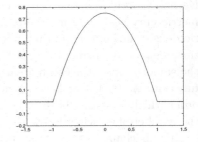

Fig. 1.2 $K(x) = \frac{3}{4}(1-x^2)I_{[-1,1]}(x)$, Epanechnikov kernel.

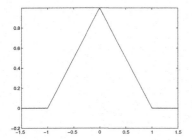

Fig. 1.3 $K(x) = (1-|x|)I_{[-1,1]}(x)$, triangle kernel.

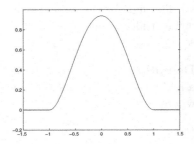

Fig. 1.4 $K(x) = \frac{15}{16}(1-x^2)^2 I_{[-1,1]}(x)$, quartic kernel.

Example 1.2. Figures 1.5 and 1.6 present some kernels from $S_{1,3}$.

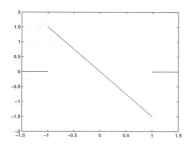

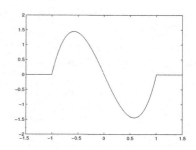

Fig. 1.5 $K(x) = -\frac{3}{2}xI_{[-1,1]}(x)$.

Fig. 1.6 $K(x) = -\frac{15}{4}x(1 - x^2) \times I_{[-1,1]}(x)$.

Example 1.3. Figures 1.7 and 1.8 present some kernels from $S_{2,4}$.

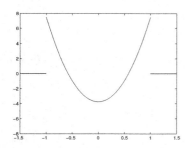

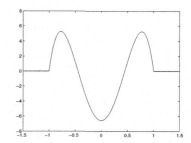

Fig. 1.7 $K(x) = -\frac{15}{4}(1 - 3x^2) \times I_{[-1,1]}(x)$.

Fig. 1.8 $K(x) = -\frac{105}{16}(1 - x^2)(1 - 5x^2)I_{[-1,1]}(x)$.

Since the dependency of the Mean (Integrated) Square Error for kernel estimates of a density, a regression function and a hazard function on the kernel is the same, the problem of a kernel choice can be discussed in general. There is only a question which optimality criteria should be adapted. For a comprehensive review on the problem of optimal kernel choice, see, *e.g.*, Granovsky and Müller (1989); Müller (1988).

We focus on a class of minimum variance kernels which minimize the asymptotic variance of kernel estimates and then optimal kernels will be treated.

Remark 1.1. Denote $V(g) = \int g^2(x)dx$ for any square integrable function, *i.e.*, $V(g)$ presents generally the squared Lebesgue's L_2 norm of a function g. This notation will be used throughout the book.

We assume that $K \in S_{\nu,k}$, $0 \le \nu \le k - 2$, and ν and k are of the same parity. In the context of the asymptotic variance of the kernel estimator the following variational problem arose

$$\min V(K), \qquad \text{subject to } K \in S_{\nu,k}. \qquad (1.2)$$

The solution of the variational problem (1.2) is called a *minimum variance kernel*.

Theorem 1.1. *Minimum variance kernels* $K \in S_{\nu,k}$ *are unique polynomials of degree* $k - 2$ *restricted on the interval* $[-1, 1]$. *These polynomials are symmetric if* k *is even and antisymmetric if* k *is odd. They have* $k - 2$ *different roots in* $(-1, 1)$. *The explicit formulas are given by means of the Legendre polynomials* P_i, $i = 0, \ldots, k - 2$

$$K(x) = \frac{(-1)^\nu \nu!}{2} \sum_{i=\nu}^{k-2} (2i + 1) p_\nu^i P_i(x),$$

where $P_i(x) = \sum_{r=0}^{i} p_r^i x^r$ *(see Complements for properties of these polynomials).*

Proof.   For proof see, *e.g.*, Müller (1988) or Horová (2000).   □

Example 1.4. The minimum variance kernel for class $S_{0,2}$ is the uniform kernel $K(x) = \frac{1}{2}I_{[-1,1]}(x)$ (see Fig. 1.1). In cases of $S_{1,3}$ and $S_{2,4}$ the minimum variance kernels are represented on Fig. 1.5 and Fig. 1.7, respectively.

Remark 1.2. Minimum variance kernels exhibit jumps at the endpoints -1, 1 of their support, which in general leads to a bad finite behavior.

It will be shown in next chapters that the dependence of Mean (Integrated) Square Error of kernel estimates on the kernel function is given through the functional

$$T(K) = \left(\left| \int_{-1}^{1} x^k K(x)dx \right|^{2\nu+1} \left(\int_{-1}^{1} K^2(x)dx \right)^{k-\nu} \right)^{\frac{2}{2k+1}}, \quad K \in S_{\nu,k},$$

$$(1.3)$$

i.e.,

$$T(K) = \left(|\beta_k|^{2\nu+1} V(K)^{k-\nu}\right)^{\frac{2}{2k+1}}.$$

Let us denote

$$\gamma_{\nu k} = \left(\frac{V(K)}{\beta_k^2}\right)^{\frac{1}{2k+1}}. \tag{1.4}$$

This parameter is called a *canonical factor* of the kernel $K \in S_{\nu,k}$.

The problem of minimizing the functional (1.3) over the set of functions $S_{\nu,k}$ has no solution (see Müller (1988)). It was therefore suggested to impose on K the following additional condition:

$$K \text{ has exactly } k - 2 \text{ sign changes in the interval } [-1, 1]. \tag{1.5}$$

(See Complements for a precise definition of sign changes of the function f.)

In next considerations, we again assume that ν, k are of the same parity (for more details see, *e.g.*, Granovsky and Müller (1989); Granovsky *et al.* (1995)). The following variational problem

$$\min T(K), \quad \text{provided that } K \text{ satisfies (1.1) and (1.5)} \tag{1.6}$$

should be solved. The solutions of the problem (1.6) are referred to *optimal kernels*.

Theorem 1.2. *Optimal kernels are polynomials of degree k. They have $k-2$ different roots in $(-1, 1)$ and $-1, 1$ are also roots of these polynomials. They are symmetric if k is even and antisymmetric if k is odd. The explicit formulas can be expressed by means of the Legendre polynomials P_i*

$$K_{opt}(x) = \frac{(-1)^\nu \nu!}{2} \sum_{i=\nu}^{k} (2i + 1) p_\nu^i (P_i(x) - P_k(x)),$$

where $P_i(x) = \sum_{r=0}^{i} p_r^i x^r$.

Proof. See, *e.g.*, Müller (1988); Granovsky *et al.* (1995). □

Example 1.5. Some examples of exact formulas for optimal kernels with their canonical factors are listed in Table 1.1, Table 1.2 and Table 1.3.

To extend our considerations we can add a request on a smoothness of the kernel function. The smoothness of some order μ is sometimes required for the kernel estimate. This can be achieved by using kernels of the corresponding smoothness, *i.e.*, $K \in S_{\nu,k} \cap C^\mu[-1, 1]$,

Table 1.1   Optimal kernels for $\nu = 0$

k	γ_{0k}	K_{opt} on $[-1, 1]$
2	1.7188	$-\frac{3}{4}(x^2 - 1)$
4	2.0165	$\frac{15}{32}(x^2 - 1)(7x^2 - 3)$
6	2.0834	$-\frac{105}{256}(x^2 - 1)(33x^4 - 30x^2 + 5)$

Table 1.2   Optimal kernels for $\nu = 1$

k	γ_{1k}	K_{opt} on $[-1, 1]$
3	1.4204	$\frac{15}{4}x(x^2 - 1)$
5	1.7656	$-\frac{105}{32}x(x^2 - 1)(9x^2 - 5)$
7	1.8931	$\frac{315}{32}x(x^2 - 1)(143x^4 - 154x^2 + 35)$

Table 1.3   Optimal kernels for $\nu = 2$

k	γ_{2k}	K_{opt} on $[-1, 1]$
4	1.3925	$-\frac{105}{16}(x^2 - 1)(5x^2 - 1)$
6	1.6964	$\frac{315}{64}(x^2 - 1)(77x^4 - 58x^2 + 5)$
8	1.8269	$-\frac{3465}{2048}(x^2 - 1)(1755x^6 - 2249x^4 + 721x^2 - 35)$

Table 1.4   Smooth optimal kernels

ν	k	μ	K_{opt} on $[-1, 1]$
0	2	1	$\frac{15}{16}(x^2 - 1)^2$
0	2	2	$-\frac{35}{32}(x^2 - 1)^3$
1	3	1	$\frac{105}{16}x(x^2 - 1)^2$

$K^{(j)}(-1) = K^{(j)}(1) = 0$, $j = 0, 1, \ldots \mu$. The class of *smooth kernels* is denoted by $S_{\nu,k}^{\mu}$. It is clear that if $K \in S_{0,k}^{\nu}$ then $K^{(\nu)} \in S_{\nu,k+\nu}^0$. The exact formula for these kernels is based on Gegenbauer polynomials and can be found, *e.g.*, in Horová (2002); Müller (1988). The algorithm for computing *smooth optimal kernels* is also implemented in MATLAB toolbox (see Sec. 1.2). The quartic kernel $K(x) = \frac{15}{16}(1 - x^2)^2 I_{[-1,1]} \in S_{0,2}^1$ is a commonly known kernel of smoothness 1, the triweight kernel $K(x) = \frac{35}{32}(1 - x^2)^3 I_{[-1,1]} \in S_{0,2}^2$ is a kernel of smoothness 2 and its second derivative $K^{(2)}(x) = -\frac{105}{16}(x^2 - 1)(5x^2 - 1)I_{[-1,1]} \in S_{2,4}^0$ is a kernel of smoothness 0. Some commonly used smooth optimal kernels are listed in Table 1.4.

Remark 1.3. In this context we emphasize that the optimal kernels belong to the class $S_{\nu,k}^0$.

The next theorem describes an interesting relation between minimum variance kernels and optimal kernels (see, *e.g.*, Horová (2000)).

Theorem 1.3. *Let* $K \in S_{\nu+1,k+1}$ *be a minimum variance kernel and* $K_{opt} \in S_{\nu,k}^0$ *be an optimal kernel. Then*

$$K'_{opt}(x) = K(x), \quad x \in [-1, 1].$$

Proof. See Complements for a sketch of the proof. □

Notation 1.1. For any $\delta > 0$ and $K \in S_{\nu,k}$ we put

$$K_\delta(x) = \frac{1}{\delta^{\nu+1}} K\left(\frac{x}{\delta}\right). \tag{1.7}$$

Lemma 1.1. *Let* $K \in S_{\nu,k}$. *The functional* $T(K)$ *is invariant with respect to the transformation*

$$K(\cdot) \to \frac{1}{\delta^{\nu+1}} K\left(\frac{\cdot}{\delta}\right) = K_\delta(\cdot), \text{ i.e., } T(K) = T(K_\delta).$$

Proof. The proof is evident. □

The kernels K, K_δ are called *equivalent kernels*.

Remark 1.4. In practical processing we encounter data which are bounded in some interval. The quality of the estimate in the boundary region is affected since the "effective" window $[x + h, x - h]$ does not belong to this interval, so the finite equivalent of the moment conditions on the kernel function does not apply any more. This phenomenon is called the *boundary effect*. There are several methods to cope with boundary effects. One of them is based on the construction of special *boundary kernels*. Their construction is described in details for instance in Müller (1991) or Horová (1997). These kernels can be used well in kernel regression but their use in density or distribution function estimates gives often inappropriate results.

Now, we remind the *convolution* notation

$$(f * g)(x) = \int f(x - y)g(y)dy,$$

for square integrable functions f and g. The next definition introduces an interesting function useful in the analysis of the performance of kernel estimations.

Definition 1.2. For any $K \in S_{0,k}$ let us define the function

$$\Lambda(z) = (K * K * K * K - 2K * K * K + K * K)(z) \qquad (1.8)$$

$$\Lambda_h(z) = \frac{1}{h}\Lambda\left(\frac{z}{h}\right) = (K_h * K_h * K_h * K_h - 2K_h * K_h * K_h + K_h * K_h)(z).$$

This function plays an important role in a special bandwidth selection method and it is a unifying link connecting kernel estimates of density function, its derivative and distribution function. It is easy to show that the following lemma is valid.

Lemma 1.2. *Let $K \in S_{0,k}$, then*

$$\int z^j \Lambda(z)dz = 0, \quad j = 0, 1, \dots, 2k - 1,$$

$$\int z^{2k} \Lambda(z)dz = \binom{2k}{k}\beta_k^2. \qquad (1.9)$$

Proof.   The proof is evident.   □

The derivatives of the function Λ take the form

$$\Lambda^{(2i)}(z) = \left(K^{(i)} * K^{(i)} * K * K - 2K^{(i)} * K^{(i)} * K + K^{(i)} * K^{(i)}\right)(z),$$

$i = 0, 1, \dots, k.$
Moreover, let us consider

$$W(z) = \int\limits_{-\infty}^{z} K(t)dt$$

and define the function

$$\Omega(z) = (W * W * K * K - 2W * W * K + W * W)(z).$$

Then

$$\Omega''(z) = \Lambda(z).$$

These facts will be used in Chap. 2 and Chap. 3.

1.2 Use of MATLAB toolbox

In the toolbox, kernel functions are created by the command K_def.

Syntax
```
K = K_def(type)
K = K_def('gauss',s)
K = K_def(method,par1,par2,par3)
```

Description
K_def creates a kernel function which satisfies conditions of the definition in Sec. 1.1.

K = K_def(type) creates a kernel as a predefined type, where type is a string variable.
Predefined types are:

'epan'	Epanechnikov kernel
'quart'	quartic kernel
'rect'	uniform (rectangular) kernel
'trian'	triangular kernel
'gauss'	Gaussian kernel

K = K_def('gauss',s) creates the Gaussian kernel with variance s^2.

K = K_def(method,par1,par2,par3) creates a kernel by a specified method (string) with parameters par1, par2, par3 (string or double). All possible combinations of method and parameters are listed in Table 1.5.

Output
The output variable K is a structure array with fields described in Table 1.6.

Table 1.5　Combinations of parameters for K_def

method	parameters	purpose
'opt'	double values ν, k, μ	optimal kernel from $S_{\nu,k}$ of smoothness μ
'str'	par1 a string formula (in variable 'x'), par2, par3 double values	kernel defined by the formula par1 normalized on the support [par2, par3]
'pol'	par1 double vector, par2, par3 double values	polynomial kernel defined by the vector of coefficients par1 normalized on the support [par2, par3]
'fun'	par1 a string, par2, par3 double values	kernel defined by the external function par1 normalized on the support [par2, par3]

Table 1.6　Structure of output

field	description
type	type of the kernel
	'opt'　　optimal kernel (default)
	'str'　　string (variable denoted as 'x')
	'pol'　　polynomial kernel
	'fun'　　external function
	'tri'　　triangular kernel
	'gau'　　Gaussian kernel
name	string with kernel expression or function name, ignored for optimal and polynomial kernel
coef	coefficients of the optimal or polynomial kernel
support	support of the kernel, default $[-1, 1]$
nu, k, mu	order and smoothness of the kernel
var	variance of the kernel $V(K)$
beta	k-th moment of the kernel β_k

Example 1.6. (Gaussian kernel)
K = K_def('gauss') gives

```
K =        type:    'gau'
           name:    ' '
           coef:    1
        support:    [-Inf, Inf]
             nu:    0
              k:    2
             mu:    Inf
            var:    0.2821
           beta:    1
```

Example 1.7. (the Epanechnikov kernel)
In this case, we have two possibilities how to create the kernel. K =
K_def('opt',0,2,0) or K = K_def('epan') gives

```
K =        type:    'opt'
           name:    ' '
           coef:    [-3/4 0 3/4]
        support:    [-1, 1]
             nu:    0
              k:    2
             mu:    0
            var:    3/5
           beta:    1/5
```

For evaluation of the kernel K in a vector x use the function K_val with
syntax value = K_val(K,x).

1.3 Complements

Let us introduce some properties of Legendre polynomials P_n (see, *e.g.*,
Szegő (1939)):

(i) Orthogonality on the interval $[-1, 1]$

$$\int_{-1}^{1} P_n(x)P_m(x)dx = \begin{cases} 0, & n \neq m, \\ \frac{2}{2n+1}, & n = m. \end{cases}$$

(ii) Recursive formula

$$P_{n+1} = \frac{2n+1}{n+1}xP_n(x) - \frac{n}{n+1}P_{n-1}(x), \qquad n = 1, 2, \ldots$$

with $P_0(x) = 1$ and $P_1(x) = x$.

(iii) Rodrigues' formula

$$P_n(x) = \frac{(-1)^n}{2^n n!} \frac{d^n}{dx^n} (1 - x^2)^n.$$

(iv) P_n is a solution of the differential equation

$$(1 - x^2)y'' - 2xy' + n(n + 1)y = 0.$$

(v) P_n satisfies

$$P_n(x) = \frac{1}{2n + 1} \left(P'_{n+1}(x) - P'_{n-1}(x) \right).$$

(vi) Boundary condition $P_n(-1) = (-1)^n P_n(1)$.

Definition 1.3. A real valued function f defined on a finite or infinite interval $[a, b]$ is said to have r *changes of sign* on $[a, b]$ if there are $r + 1$ intervals $[x_{i-1}, x_i] \subset [a, b]$, $i = 1, \ldots, r + 1$, $x_0 = a$, $x_{r+1} = b$, such that

(i) $f(x)f(y) \geq 0$ for all $x, y \in [x_{i-1}, x_i]$, $i = 1, \ldots, r + 1$ with strict inequality for all x, y from a subset $D \subset [x_{i-1}, x_i]$ of nonzero Lebesgue measure

(ii) $f(x)f(y) \leq 0$ for all $x \in [x_{i-1}, x_i], y \in [x_i, x_{i+1}], i = 1, \ldots, r$.

Denote $ch(f) = r$ the number of changes of sign of the function f on $[a, b]$.

In Mammitzsch (1985), the following result has been proved:

Let $K \in S_{\nu,k}$. Then $ch(K) \geq k - 2$.

Sketch of proof of Theorem 1.3 (Horová (1996)):
Using the property (v) of Legendre polynomials gives

$$K'_{opt}(x) = \frac{(-1)^\nu \nu!}{2} \sum_{i=\nu}^{k} (2i + 1)p_\nu^i (P'_i(x) - P'_k(x)),$$

where

$$P'_i(x) - P'_k(x) = -\sum_{j=1}^{\frac{k-i}{2}} (2(i + 2j - 1) + 1)P_{i+2j-1}(x)$$

($\frac{k-i}{2}$ is an integer because k, i are of the same parity).
Thus

$$K'_{opt}(x) = \frac{(-1)^{\nu+1}\nu!}{2} \sum_{i=\nu}^{k} (2i+1)p_\nu^i \sum_{j=1}^{\frac{k-i}{2}} (2(i+2j-1)+1)P_{i+2j-1}(x). \quad (1.10)$$

Recall that the minimum variance kernel $K \in S_{\nu+1,k+1}$ takes the form

$$K(x) = \frac{(-1)^{\nu+1}(\nu+1)!}{2} \sum_{i=\nu+1}^{k-1} (2i+1)p_{\nu+1}^i P_i(x). \quad (1.11)$$

Easy calculation shows that

$$(\nu+1)p_{\nu+1}^{2j+1+\nu} = \sum_{i=\nu}^{2j+\nu} (2i+1)p_\nu^i. \quad (1.12)$$

Now by comparing coefficients standing by $P_{\nu+2j+1}(x)$ in (1.10), (1.11) and taking (1.12) into account we arrive at the relation

$$K'_{opt}(x) = K(x).$$

Chapter 2

Univariate kernel density estimation

The probability density function is a fundamental concept in statistics. Density estimation is the reconstruction of the density function from a set of observed data and it can provide useful information of the given data set.

Suppose that $X_1, \ldots, X_n$ is a set of continuous random variables having common density f. One approach of estimating f is parametric. Assume that data chosen from one of a known parametric density family depending on some parameters and the estimate consists in obtaining the estimates of these parameters. Parametric curves rely on model building and prior knowledge of the equations underlying data (Fisher (1922, 1932); Scott (1992)).

In the nonparametric case, the emphasis is directly on obtaining a good estimate of the entire density f. Nonparametric estimates attempt to construct the probability density from which the sample has come using sample values and as few assumptions about the density as possible. Nonparametric curves are driven by structure in data and are broadly applicable.

The oldest nonparametric density estimator is the histogram (Silverman (1986); Scott (1992); Wand and Jones (1995)). Although the idea of grouping data in the form of a histogram is at least as old as Graunt's work in 1662 (Graunt (1662)), no systematic guidelines for designing histograms were given until Herber Sturges' short note (Sturges (1926)). His work made use of a device that was advocated more generally by Tukey (Tukey (1977)). Though the histogram remains an excellent tool for data presentation, it is worth at least considering alternative density estimates that are available.

Substantial and simultaneous progress was made in nonparametric density estimates after the 1950s. Rosenblatt (1956); Whittle (1958); Parzen

(1962) developed more general kernel estimators. These papers were followed by many theoretical papers and important monographs: Silverman (1986); Scott (1992); Härdle *et al.* (2004); Bowman and Azzalini (1997). Apart from the histogram, the kernel estimator is probably the most commonly used estimator and its properties have been studied by many authors.

2.1 Basic definition

Let $X_1, \ldots, X_n$ be independent real random variables having the same continuous density f. The symbol $\hat{f}$ will be used to denote whatever density estimation is currently being considered.

The *histogram* is a widely used nonparametric density estimator. Given an origin x_0 and a bin width h, we define the bins of the histogram to be the intervals $[x_0 + mh, x_0 + (m+1)h]$ for integers m. The intervals have been chosen closed on the left and open on the right for definiteness. The histogram is then defined as

$$\hat{f}(x, h) = \frac{1}{nh}(\text{number of } X_i \text{ in the same bin as } x). \qquad (2.1)$$

Two choices are to be made when constructing the histogram: an origin x_0 and a bin width h. The histogram is very sensitive to the placement of the origin, *i.e.*, on bin edges. The bin edge problem is one of the histogram's main disadvantages. Moreover, the underlying density is usually assumed to be smooth, but the histogram is a step function.

Example 2.1. We simulated a data sample having the density f in the form of a mixture of two normal densities (see Fig. 2.1)

$$f(x) = 0.3 \frac{1}{\sqrt{2\pi\sigma_1^2}} e^{-\frac{(x-\mu_1)^2}{2\sigma_1^2}} + 0.7 \frac{1}{\sqrt{2\pi\sigma_2^2}} e^{-\frac{(x-\mu_2)^2}{2\sigma_2^2}}$$

for $\mu_1 = 0$, $\mu_2 = 2$, $\sigma_1 = \sigma_2 = \frac{1}{2}$ and $n = 200$. The histogram of data is presented by Fig. 2.2.

Rosenblatt (1956); Whittle (1958); Parzen (1962) developed an approach which removes the aforementioned difficulties. This approach is based on the following steps:

- a smooth kernel function rather than a box is used as the basic building block,
- these smooth functions are centered directly over each observation.

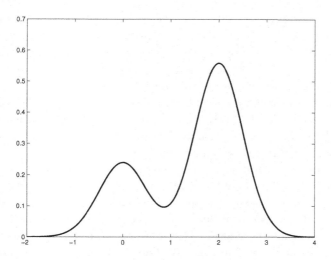

Fig. 2.1   Mixture of two normal densities.

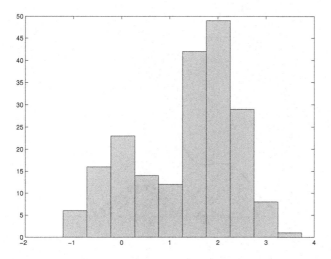

Fig. 2.2   Histogram for simulated data from example 2.1.

Thus, the kernel estimator (Rosenblatt-Parzen) of f at the point $x \in \mathbb{R}$ is defined as

$$\hat{f}(x,h) = \frac{1}{nh} \sum_{i=1}^{n} K\left(\frac{x - X_i}{h}\right) = \frac{1}{n} \sum_{i=1}^{n} K_h(x - X_i), \qquad (2.2)$$

where $K_h(t) = \frac{1}{h} K\left(\frac{t}{h}\right)$, $K \in S_{0,k}$, $h > 0$. The positive number h is a smoothing parameter called also a *bandwidth*. The bandwidth h is depending on n, $h = h(n)$: $\{h(n)\}_{n=1}^{\infty}$ is a nonrandom sequence of positive numbers. To keep the notation less cumbersome the dependence of h on n will be suppressed in our next considerations.

The kernel estimates depend on three parameters: the kernel K which plays the role of a weight function, the bandwidth which controls the smoothness of the estimate and the order of the kernel which is related to number of derivatives assumed to exist in the model.

The estimate $\hat{f}$ will inherit all the continuity and differentiability properties of K. Thus the smooth K will produce a nice looking curve.

The problem of choosing the smoothing parameter is of a crucial importance and will be treated in Sec. 2.3. Since visualization is an important component of a nonparametric data analysis we clarify the construction of pointwise estimate of f in Fig. 2.3. This figure shows individual kernels

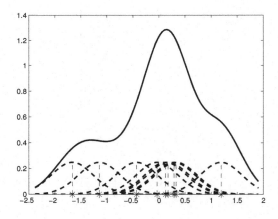

Fig. 2.3   Construction of the kernel density estimate.

and the resulting estimate $\hat{f}$. The kernel estimate is constructed by centering a scaled kernel K_h at each X_i, $i = 1, \ldots, n$ and $\hat{f}(x)$ is the average of n kernel ordinates at the point x.

2.2 Statistical properties of the estimate

The problem we are going to discuss is the closeness of the estimate $\hat{f}$ to the true density f in various senses. Our analysis requires the specification of an appropriate error criterion for measuring the error when estimating the density at a single point as well as the error when estimating the density over the whole real line. A useful criterion when estimating at a single point is the Mean Square Error (MSE) defined by

$$\text{MSE}\{\hat{f}(x,h)\} = E\{\hat{f}(x,h) - f(x)\}^2.$$

It is very easy to show that the following lemma is valid.

Lemma 2.1.

$$\text{MSE}\{\hat{f}(x,h)\} = (E\hat{f}(x,h) - f(x))^2 + E(\hat{f}(x,h))^2 - E^2(\hat{f}(x,h)).$$

Remark 2.1. The term $E\hat{f}(x,h) - f(x)$ is called bias$\{\hat{f}(x,h)\}$ and $E(\hat{f}(x,h))^2 - E^2(\hat{f}(x,h))$ is known as var$\{\hat{f}(x,h)\}$. So the previous lemma describes the decomposition of MSE into a variance and a squared bias. This decomposition provides an easier analysis and interpretation of the performance of the kernel density estimator.

The convolution notation allows to express MSE$\{\hat{f}(x,h)\}$ in the following form.

Lemma 2.2.

$$\text{MSE}\{\hat{f}(x,h)\} = \underbrace{\{(K_h * f)(x) - f(x)\}^2}_{\text{bias}^2\{\hat{f}(x,h)\}}$$

$$+ \underbrace{\frac{1}{n}\{(K_h^2 * f)(x) - (K_h * f)^2(x)\}}_{\text{var}\{\hat{f}(x,h)\}}.$$

As concerns a global criterion we consider the Integrated Square Error (ISE) given by

$$\text{ISE}\{\hat{f}(\cdot,h)\} = \int \{\hat{f}(x,h) - f(x)\}^2 dx.$$

However it will be more appropriate to analyze the expected value of this random quantity, the Mean Integrated Square Error (MISE)

$$\text{MISE}\{\hat{f}(\cdot,h)\} = E \int \{\hat{f}(x,h) - f(x)\}^2 dx.$$

Remark 2.2. By changing the order of integration we obtain

$$\text{MISE}\{\hat{f}(\cdot,h)\} = \int \text{MSE}\{\hat{f}(x,h)\} dx.$$

Making some calculations we arrive at more suitable expression.

Lemma 2.3.

$$\text{MISE}\{\hat{f}(\cdot, h)\} = \frac{1}{nh} \int K^2(x)dx$$

$$+ (1 - n^{-1}) \int (K_h * f)^2(x)dx$$

$$- 2 \int (K_h * f)(x)f(x)dx + \int f^2(x)dx.$$

A problem with MSE and MISE expressions is that they depend on the bandwidth in a complicated way. This makes it difficult to interpret the influence of the bandwidth on the performance of the kernel density estimator. In this section, we investigate one way of overcoming this problem that involves the derivation of large sample approximations for leading variance and bias terms. These approximations have very simple expressions that allow a deeper appreciation of the role of the bandwidth. They can also be used to obtain the rate of convergence of the kernel density estimator and the MISE-optimal bandwidth

$$h_{\text{MISE}} = \arg\min \text{MISE}\{\hat{f}(\cdot, h)\}.$$

Theorem 2.1. *Let* $f \in C^{k_0}$, $0 < k \le k_0$, *k even,* $f^{(k)}$ *be square integrable,* $K \in S_{0,k}$, $\lim_{n \to \infty} h = 0$, $\lim_{n \to \infty} nh = \infty$. *Then*

$$\text{MISE}\{\hat{f}(\cdot, h)\} = \frac{1}{nh}V(K) + h^{2k}\beta_k^2 D_k + o\left\{h^{2k} + (nh)^{-1}\right\}, \qquad (2.3)$$

where $D_k = \int \left(\frac{f^{(k)}(x)}{k!}\right)^2 dx.$

Proof. Proof of this theorem for $k = 2$ can be found, *e.g.*, in Wand and Jones (1995); Scott (1992); Silverman (1986). □

Corollary 2.1. *Under the given assumptions* $\hat{f}(\cdot, h)$ *is a consistent estimate of* f.

Since MISE is not mathematically tractable we employ the Asymptotic Mean Integrated Square Error (AMISE) which can be written as a sum of the Asymptotic Integrated Variance and the Asymptotic Integrated Square Bias

$$\text{AMISE}\{\hat{f}(\cdot, h)\} = \underbrace{\frac{V(K)}{nh}}_{\text{AIV}} + \underbrace{h^{2k}\beta_k^2 D_k}_{\text{AISB}}. \qquad (2.4)$$

Clearly, the last expression highlights the own role of each of three parameters h, k, K and also their interactions.

Let $K \in S_{0,k}$ and let us recall the notation (1.7) from Chap. 1. For any $\delta > 0$

$$K_\delta(x) = \frac{1}{\delta} K\left(\frac{x}{\delta}\right). \tag{2.5}$$

It is easy to find out that the amount of smoothing given by the kernel K with the bandwidth h is the same as that given by the kernel K_δ with the bandwidth $h^* = h/\delta$. In Chap. 1, the canonical factor γ_{0k} for $K \in S_{0,k}$ was defined by formula (1.4). In this context the kernel $K_{\gamma_{0k}}$ is called a *canonical kernel*.It should be emphasized that the canonical kernel $K_{\gamma_{0k}}$ possesses a very useful property: $V(K_{\gamma_{0k}}) = \beta_k^2(K_{\gamma_{0k}})$.

Remark 2.3. Evidently, AMISE of the estimate $\hat{f}$ with $K_{\gamma_{0k}}$ and h^* is of the form

$$\text{AMISE}\{\hat{f}(\cdot, h^*)\} = \frac{V(K_{\gamma_{0k}})}{nh^*} + h^{*2k}\beta_k^2(K_{\gamma_{0k}})D_k$$

and with respect to the fact mentioned above the contribution of the kernel $K_{\gamma_{0k}}$ to both parts of $\text{AMISE}\{\hat{f}(\cdot, h^*)\}$ is the same.

Further, in terms of the functional $T(K)$, AMISE can be expressed in the following way (Horová *et al.* (2002); Marron and Nolan (1988)):

$$\text{AMISE}\{\hat{f}(\cdot, h)\} = T(K)\left(\frac{\gamma_{0k}}{nh} + \frac{h^{2k}D_k}{\gamma_{0k}^{2k}}\right), \tag{2.6}$$

where the functional $T(K)$ is defined by (1.3) with $\nu = 0$.

Now, minimize (2.6) with respect to h and put

$$h_{opt,0,k} = \arg\min_{h \in H_n} \text{AMISE}\{\hat{f}(\cdot, h)\} = h_{\text{AMISE}},$$

where H_n is the set of acceptable bandwidths.

In order to obtain $h_{opt,0,k}$ we solve the equation

$$\frac{d\text{AMISE}\{\hat{f}(\cdot, h)\}}{dh} = 0 \tag{2.7}$$

and the calculation gives

$$h_{opt,0,k}^{2k+1} = \frac{\gamma_{0k}^{2k+1}}{2nkD_k}, \tag{2.8}$$

i.e., $h_{opt,0,k} = O(n^{-\frac{1}{2k+1}})$.

Moreover,

$$\frac{d^2\text{AMISE}\{\hat{f}(\cdot, h)\}}{dh^2}\Big|_{h=h_{opt,0,k}} = O\left(n^{-\frac{2k-2}{2k+1}}\right).$$

Thus we can state that for higher order kernels the minimum of $\text{AMISE}\{\hat{f}(\cdot, h)\}$ in $h_{opt,0,k}$ gets flatter (see Fig. 2.4, solid, dashed and dotted lines have been obtained for $k = 2, 4, 6$, successively) and thus a bandwidth choice close to the optimal one does not lead to a large increase in $\text{AMISE}\{\hat{f}(\cdot, h)\}$. Let us notice that for $k = 2$ the optimal bandwidth

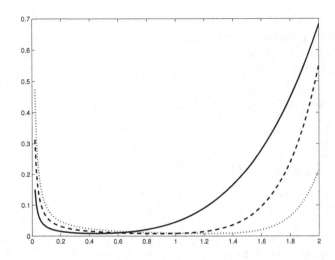

Fig. 2.4 $\text{AMISE}\{\hat{f}(\cdot, h)\}$ for data from Example 2.1 and $k = 2, 4, 6$.

$h_{opt,0,k}$ also realizes the minimum of $\frac{d^2\text{AMISE}\{\hat{f}(\cdot, h)\}}{dh^2}$.

Substitution $h_{opt,0,k}$ into (2.6) yields

$$\text{AMISE}\{\hat{f}(\cdot, h_{opt,0,k})\} = T(K) n^{-\frac{2k}{2k+1}} D_k^{\frac{1}{2k+1}} \frac{2k+1}{(2k)^{\frac{2k}{2k+1}}}.$$

It follows that $\text{AMISE}\{\hat{f}(\cdot, h_{opt,0,k})\} = O\left(n^{-\frac{2k}{2k+1}}\right)$.

Let us recall two parts of $\text{AMISE}\{\hat{f}(\cdot, h)\}$ defined in (2.4)

$$\text{AIV}\{\hat{f}(\cdot, h)\} = \frac{V(K)}{nh} = \frac{T(K)\gamma_{0k}}{nh} \tag{2.9}$$

and

$$\text{AISB}\{\hat{f}(\cdot, h)\} = h^{2k}\beta_k^2 D_k = \frac{T(K)D_k h^{2k}}{\gamma_{0k}^{2k}}. \tag{2.10}$$

The optimal bandwidth is the solution of (2.7). Moreover, it is easy to see that the following lemma is valid.

Lemma 2.4.

$$\text{AIV}\{\hat{f}(\cdot, h_{opt,0,k})\} - 2k\text{AISB}\{\hat{f}(\cdot, h_{opt,0,k})\} = 0. \tag{2.11}$$

This relation is of a great importance since it serves as a basis of one of data-driven bandwidth selectors methods.

The parameters K, h, k are closely related one to each other and thus it is necessary to address them all together for a suitable procedure. The calculation D_k from (2.8) and substituting it into (2.6) yields an appropriate formula in Theorem 2.2.

Theorem 2.2.

$$\text{AMISE}\{\hat{f}(\cdot, h_{opt,0,k})\} = T(K)\frac{(2k+1)\gamma_{0k}}{2nkh_{opt,0,k}}. \tag{2.12}$$

This formula makes possible to choose the kernel, its order, and the smoothing parameter simultaneously in some automatic, and optimal way (see Horová *et al.* (2002)). The automatic procedure will be described in Sec. 2.6.

2.3 Choosing the shape of the kernel

Choice of the kernel does not influence the asymptotic behavior of the estimate so significantly as the bandwidth does. Thus we restrict ourselves only on two classes of useful kernels.

The Asymptotic Integrated Variance AIV of the kernel estimate $\hat{f}$ depends on $V(K)$. Thus to minimize AIV with respect to K, the minimum variance kernel is suitable (Theorem 1.1). The separation between kernel and bandwidth effects (Theorem 2.2) suggested naturally looking for a kernel that minimizes the quantity $T(K)$. These kernels were discussed in Chap. 1 and their formulas are given in Theorem 1.2.

A clear advantage of optimal kernels which minimize the functional $T(K)$ is given by the fact that they are continuous on $\mathbb{R}$ and the smoothness of K is inherited by the estimate $\hat{f}$.

Generally, minimum variance kernels exhibit jumps at the end points -1, 1 of their support, which in general leads to a bad finite sample behavior. See Fig. 2.5 and Fig. 2.6 to compare kernel density estimates over the data set from Example 2.1. The first figure illustrates the case of using the minimum variance kernel $K(x) = \frac{1}{2}I_{[-1,1]}$ (uniform kernel) and the second one illustrates the case of using the optimal kernel $K(x) = \frac{3}{4}(1 - x^2)I_{[-1,1]}$ (the Epanechnikov kernel).

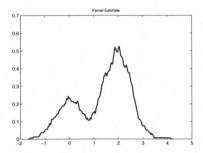
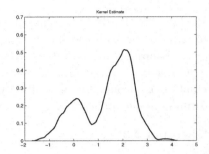

Fig. 2.5   Kernel density estimate with the uniform kernel.

Fig. 2.6   Kernel density estimate with the Epanechnikov kernel.

2.4   Choosing the bandwidth

The problem of choosing how much to smooth, *i.e.*, how to choose the bandwidth is a crucial common problem in kernel smoothing.

Methods for a bandwidth choice have been developed in many papers and monographs, see, *e.g.*, Cao *et al.* (1994); Chaudhuri and Marron (1999); Härdle *et al.* (2004); Horová and Zelinka (2007a); Silverman (1986); Wand and Jones (1995); Scott (1992); Fan and Gijbels (1995), and many others.

However there does not exist any universally accepted approach to this serious problem yet. The appropriate choice of a bandwidth has been always influenced by the purpose for which the kernel estimate is used (see Marron (1996) for a discussion). This idea was nicely characterized in Silverman (1986), p. 44:

"For many purposes, particularly for model and hypothesis generation, it is by no means unhelpful for the statistician to supply the scientist with a range of possible presentation of the data."

The simplest method of bandwidth choice is judging by eye. It is a good idea to plot out several curves that have been obtained with different bandwidths before embarking on a more sophisticated and automatic bandwidth choice.

It should be emphasized that, from an exploratory point of view, all choices of the bandwidth lead to useful density estimators. Large bandwidths provide a picture of the global structure in the unknown density, small bandwidths, on the other hand, reveal local structures which may or may not be present in the true density.

This approach can be illustrated by a short simulation study. For standard normal density we generate the sample of size $n = 100$. A family

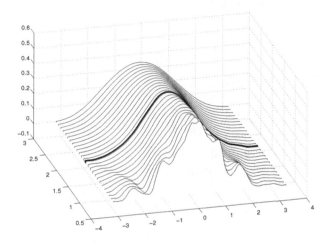

Fig. 2.7   Choosing the bandwidth.

of estimates $\{\hat{f}(x, h) | \ h \in [0.5, 3]\}$, $x \in [-3.2, 3.2]$ indexed by the bandwidth is overlaid in Fig. 2.7 as thin solid lines. The bold line denotes the estimate with the optimal bandwidth. The family shows a very wide range of smoothing, from nearby the raw data to the oversmoothed estimate (Chaudhuri and Marron (1999)).

2.4.1   *Reference rule*

The formula (2.8) provides a simple insight into an "optimal" bandwidth. But the optimal bandwidth depends on the density to be estimated. The idea of Deheuvels (1977) popularized by Silverman (1986) consists in replacing the unknown part D_k of $h_{opt,0,k}$ by an estimated value based on a parametric family. Since scale is very important for bandwidth choice, but location is not, a natural choice for parametric family is $N(0, \sigma^2)$ (Silverman (1986); Řezáč (2007)). This method is called a method of the normal reference rule (see also Scott (1992); Wand and Jones (1995)).

Let f be a standard normal density with variance σ^2. It can be shown (see, *e.g.*, Řezáč (2007)) that for $K \in S_{0,k}$

$$h_{\text{REF}} = \left(\frac{2^{2k}(k!)^3 \sqrt{\pi} V(K)}{(2k)! \beta_k^2 k} \right)^{\frac{1}{2k+1}} \sigma n^{-\frac{1}{2k+1}}$$

is an estimate of $h_{opt,0,k}$. For the Gaussian kernel the formula takes the form (Scott (1992); Silverman (1986))

$$h_{\text{REF}} = \left(\frac{4}{3n} \right)^{\frac{1}{5}} \sigma$$

and we can consider

$$\hat{h}_{\text{REF}} = \left(\frac{2^{2k}(k!)^3 \sqrt{\pi} V(K)}{(2k)! \beta_k^2 k} \right)^{\frac{1}{2k+1}} \hat{\sigma} n^{-\frac{1}{2k+1}}$$

as the estimate of the optimal bandwidth.
Common choices of $\hat{\sigma}$ are the sample standard deviation

$$\hat{\sigma}_{SD} = \left(\frac{1}{n-1} \sum_{i=1}^{n} (\bar{X} - X_i)^2 \right)^{1/2} \tag{2.13}$$

or the standardized interquartile range

$$\hat{\sigma}_{IQR} = \frac{X_{[3n/4]} - X_{[n/4]}}{\Phi^{-1}(\frac{3}{4}) - \Phi^{-1}(\frac{1}{4})}, \tag{2.14}$$

where Φ^{-1} is the standard normal quantile function.

2.4.2 *Maximal smoothing principle*

An interesting variation of this idea is oversmoothing or maximal smoothing principle proposed by Terrell and Scott (1985); Terrell (1990) and Horová and Zelinka (2007a). They solved the variational problem of minimizing $\int \left(f^{(k)}(x) \right)^2 dx$ subject to various types of scale constraints. This solution, together with a scale estimate, results in a "maximal possible amount of smoothing" and an "oversmoothed bandwidth" that comes from using this in $h_{opt,0,k}$. Upper bounds on a reasonable amount of smoothing are quite useful in certain contexts; namely as a starting point for cross-validation methods and the iterative method which will be treated later.

Generally, the $Beta(k + 2, k + 2)$ family minimizes $\int \left(f^{(k)}(x) \right)^2 dx$ for a given standard deviation, namely the following theorem holds (Terrell (1990)).

Theorem 2.3. *Among those densities f supported on the interval $[-1, 1]$, the $Beta(k + 2, k + 2)$ density*

$$g_k(x) = \begin{cases} \frac{(2k+3)!}{(k+1)!^2 2^{2k+3}} (1 - x^2)^{k+1}, & |x| \leq 1, \\ 0, & otherwise, \end{cases}$$

has the smallest value of $\int\limits_{-1}^{1} \left(f^{(k)}(x) \right)^2 dx$.

We remind the useful facts:

(I) $\int \left((rg(rx))^{(k)} \right)^2 dx = r^{2k+1} \int \left(g^{(k)}(x) \right)^2 dx$, $r > 0$ for any density g for which this expression makes sense.

(II) If the density g has the variance σ_g^2, then the density $\frac{\sigma}{\sigma_g} g(\frac{\sigma}{\sigma_g})$ has the variance σ^2.

Since $\sigma_k^2 = \int\limits_{-1}^{1} x^2 g_k(x) dx = 1/(2k + 5)$, then (see, *e.g.*, Terrell (1990) and Horová and Zelinka (2007a))

$$h_{opt,0,k} \leq \gamma_{0,k} \left(\frac{(k!)^2}{2nk} \right)^{\frac{1}{2k+1}} \frac{\sigma}{\sigma_k} \left(\int_{-1}^{1} \left(g_k^{(k)}(x) \right)^2 dx \right)^{-\frac{1}{2k+1}}. \tag{2.15}$$

The unknown standard deviation σ can be estimated by the formulas (2.13) and (2.14). Further

$$\int\limits_{-1}^{1} \left(g_k^{(k)}(x) \right)^2 dx = \frac{1}{2^{2k+2}} \frac{\Gamma(2k + 4)\Gamma(2k + 3)}{(2k + 1)(2k + 5)(\Gamma(k + 2))^2},$$

(the proof can be found in Horová and Zelinka (2007a)).
Now, using these facts we arrive at the suitable upper bound for the optimal bandwidth $h_{opt,0,k} \leq \hat{h}_{MS}$

$$\hat{h}_{MS} = \hat{\sigma} b_k n^{-\frac{1}{2k+1}}, \tag{2.16}$$

where

$$b_k = 2\sqrt{2k + 5} \left(\frac{(2k + 1)(2k + 5)(k + 1)^2 \Gamma^4(k + 1)}{k\Gamma(2k + 4)\Gamma(2k + 3)} \right)^{\frac{1}{2k+1}} \gamma_{0k}.$$

Table 2.1   Some selected kernels and b_k

k	$K(x)$	kernel	b_k
2	$\frac{1}{\sqrt{2\pi}}e^{-\frac{1}{2}x^2}$	Gaussian	1.1439
2	$\frac{3}{4}(1-x^2)I_{[-1,1]}$	Epanechnikov	2.5324
2	$\frac{15}{16}(1-x^2)^2I_{[-1,1]}$	Quartic	3.0
4	$\frac{15}{32}(x^2-1)(7x^2-3)I_{[-1,1]}$		3.3175
6	$\frac{105}{256}(1-x^2)(33x^4-30x^2+5)I_{[-1,1]}$		3.9003

Table 2.1 brings some useful kernels and corresponding values b_k.

Remark 2.4. In conclusion of this paragraph the following bounds for acceptable bandwidths can be recommended:

- A lower bound h_l can be taken as the minimum distance between consecutive points X_i, $i = 1,\ldots,n$ (Devroye and Lugosi (1997)).
- An upper bound $h_u = \hat{h}_{\mathrm{MS}}$ is given by (2.16).

Thus the set of acceptable bandwidths is $H_n = [h_l, h_u]$.

2.4.3   *Cross-validation methods*

Among the earliest fully automatic and data-driven bandwidth selection methods are those based on cross-validation (CV) ideas (see, *e.g.*, Bowman (1984); Silverman (1986); Scott (1992); Wand and Jones (1995)). All CV bandwidth selectors aim to estimate MISE or AMISE.

The least squares cross-validation (LSCV) targets MISE and employs the objective function

$$\mathrm{LSCV}(h) = \int \hat{f}^2(x,h)dx - 2n^{-1}\sum_{i=1}^{n}\hat{f}_{-i}(X_i,h), \qquad (2.17)$$

where

$$\hat{f}_{-i}(X_i,h) = (n-1)^{-1}\sum_{\substack{j=1\\j\neq i}}^{n}K_h(X_i-X_j)$$

is the density estimate based on the sample with X_i deleted, often called a "leave-one-out" density estimator.

Remark 2.5. Notice the following interesting property for $\hat{f}_{-i}(X_i,h)$

$$\hat{f}_{-i}(X_i,h) = \frac{n}{n-1}\hat{f}(X_i,h) - \frac{K_h(0)}{n-1}.$$

It can be easy shown that $\text{LSCV}(h)$ is an unbiased estimator of the quantity $\text{MISE}\{\hat{f}(\cdot, h)\} - \int f^2(x)dx$, since

$$E\{\text{LSCV}(h)\} = \text{MISE}\{\hat{f}(\cdot, h)\} - \int f^2(x)dx.$$

Let $\hat{h}_{\text{LSCV}}$ stand for minimization of $\text{LSCV}(h)$:

$$\hat{h}_{\text{LSCV}} = \underset{h \in H_n}{\arg\min} \; \text{LSCV}(h).$$

Biased cross-validation (BCV) involves estimation of AMISE (Scott and Terrell (1987)). The BCV method employs the objective function

$$\text{BCV}(h) = (nh)^{-1}V(K) + \frac{1}{(k!)^2}h^{2k}\beta_k^2(K)\widetilde{V}(f^{(k)}), \qquad (2.18)$$

where

$$\widetilde{V}(f^{(k)}) = \frac{1}{n^2}\sum_{\substack{i,j \\ i \neq j}}^{n}(K_h^{(k)} * K_h^{(k)})(X_i - X_j)$$

is an estimate of $V(f^{(k)}) = \int (f^{(k)}(x))^2 dx$.

The BCV bandwidth selector is denoted by $\hat{h}_{\text{BCV}}$

$$\hat{h}_{\text{BCV}} = \underset{h \in H_n}{\arg\min} \; \text{BCV}(h).$$

The performance of bandwidth selectors can be assessed by its relative rate of convergence.

Definition 2.1. Let $\hat{h}$ be a data-driven bandwidth selector. We say that $\hat{h}$ converges to $h_{opt,0,k}$ with the *relative rate of convergence* $n^{-\alpha}$ if

$$\frac{\hat{h} - h_{opt,0,k}}{h_{opt,0,k}} = O(n^{-\alpha}).$$

The relative convergence rates of $\hat{h}$ to $h_{opt,0,k}$ (for case $k = 2$) are

$$\frac{\hat{h}_{\text{LSCV}} - h_{opt,0,2}}{h_{opt,0,2}} = O(n^{-\frac{1}{10}}), \quad \frac{\hat{h}_{\text{BCV}} - h_{opt,0,2}}{h_{opt,0,2}} = O(n^{-\frac{1}{10}})$$

(Hall and Marron (1987); Scott and Terrell (1987)). See also Jones and Kappenman (1991), where a class of data-based bandwidth selection procedures has been investigated.

2.4.4   *Plug-in method*

Plug-in (PI) bandwidth selectors are based on estimating of the unknown quantity $V(f^{(k)}) = \int \left(f^{(k)}(x) \right)^2 dx$ in the formula (2.8)

$$h_{opt,0,k}^{2k+1} = \frac{\gamma_{0k}^{2k+1}}{2nkD_k} = \frac{\gamma_{0k}^{2k+1}(k!)^2}{2nkV(f^{(k)})}.$$

Using integration by parts one obtains under sufficient smoothness assumptions on f

$$V(f^{(k)}) = (-1)^k \int f^{(2k)}(x)f(x)dx.$$

It is therefore sufficient to study estimation of

$$\psi_k = \int f^{(2k)}(x)f(x)dx.$$

Note that

$$\psi_k = E\{f^{(2k)}(X)\}.$$

The quantity ψ_k can be estimated by the kernel estimator $\widehat{\psi}_k(g)$ with a bandwidth g and a kernel L

$$\widehat{\psi}_k(g) = \frac{1}{n^2} \sum_{i,j=1}^{n} L_g^{(2k)}(X_i - X_j),$$

where the kernel L and the bandwidth g may be different from K and h (Hall and Marron (1987); Jones and Sheather (1991)). Then

$$\hat{h}_{\mathrm{PI}} = \left(\frac{\gamma_{0k}^{2k+1}(k!)^2}{2nk(-1)^k\widehat{\psi}_k(g)} \right)^{\frac{1}{2k+1}}. \tag{2.19}$$

This rule is not fully automatic since $\hat{h}_{\mathrm{PI}}$ depends on the choice of the pilot bandwidth g. This method is described in details, *e.g.*, in Wand and Jones (1995).

The relative rate of convergence (for $k = 2$) is

$$\frac{\hat{h}_{\mathrm{PI}} - h_{opt,0,2}}{h_{opt,0,2}} = O(n^{-\frac{5}{14}}).$$

2.4.5 *Iterative method*

The method we are going to present is based on Lemma 2.4 and on a suitable estimation of AMISE (see, *e.g.*, Müller and Wang (1990a) and Jones *et al.* (1991)). The equation (2.11) can be rewritten as

$$\frac{V(K)}{nh} - 2kh^{2k}\frac{\beta_k^2}{(k!)^2}V(f^{(k)}) = 0 \tag{2.20}$$

and minimization problem $\arg\min\limits_{h \in H_n} \text{AMISE}(h)$ is equivalent to solving this equation.

Consider estimates of a variance and a bias as follows:

$$\widehat{\text{var}}\{\hat{f}(x,h)\} = \frac{1}{nh}\int K^2(y)\hat{f}(x-hy,h)dy.$$

$$\overline{\text{bias}}\{\hat{f}(x,h)\} = (K_h * \hat{f})(x,h) - \hat{f}(x,h)$$
$$= \int \hat{f}(x-hy,h)K(y)dy - \hat{f}(x,h).$$

It is easy to see that

$$\widehat{\text{AIV}}\{\hat{f}(\cdot,h)\} = \frac{1}{nh}V(K) \tag{2.21}$$

and using convolution leads to the formula for the estimated bias (see also Definition 1.2):

$$\overline{\text{AISB}}\{\hat{f}(\cdot,h)\}$$
$$= \int \overline{\text{bias}}^2\{\hat{f}(x,h)\}dx$$
$$= \frac{1}{n^2h}\sum_{i,j=1}^{n}(K*K*K*K - 2K*K*K + K*K)\left(\frac{X_i - X_j}{h}\right)$$
$$= \frac{1}{n^2h}\sum_{i,j=1}^{n}\Lambda\left(\frac{X_i - X_j}{h}\right). \tag{2.22}$$

To avoid the fact that $\overline{\text{AISB}}$ is biased, considered unbiased estimate of AISB is

$$\widehat{\text{AISB}}\{\hat{f}(\cdot,h)\} = \frac{1}{n^2h}\sum_{\substack{i,j=1 \\ i \neq j}}^{n}\Lambda\left(\frac{X_i - X_j}{h}\right).$$

Then the estimate of AMISE is

$$\widehat{\text{AMISE}}\{\hat{f}(\cdot, h)\} = \frac{1}{nh}V(K) + \frac{1}{n^2}\sum_{\substack{i,j=1 \\ i \neq j}}^{n} \Lambda_h(X_i - X_j). \qquad (2.23)$$

Let $\hat{h}_{\text{IT},0,k}$ be the minimizing bandwidth,

$$\hat{h}_{\text{IT}} = \underset{h \in H_n}{\arg\min}\ \widehat{\text{AMISE}}\{\hat{f}(\cdot, h)\}.$$

According to the relation (2.11) the minimization of $\widehat{\text{AMISE}}$ is equivalent to solving

$$\frac{V(K)}{nh} - \frac{2k}{n^2}\sum_{\substack{i,j=1 \\ i \neq j}}^{n} \Lambda_h(X_i - X_j) = 0. \qquad (2.24)$$

In the paper Horová and Zelinka (2007a) this nonlinear equation was solved by Steffensen's method. But this equation can be rewritten as

$$\frac{2k}{n}\sum_{\substack{i,j=1 \\ i \neq j}}^{n} \Lambda\left(\frac{X_i - X_j}{h}\right) - V(K) = 0. \qquad (2.25)$$

Since the first derivative of the function standing on the left hand side of this equation is easy to compute using convolutions, Newton's method can be used with initial approximation obtained by the maximal smoothing principle.

The solution $\hat{h}_{\text{IT},0,k}$ of the equation (2.24) or (2.25) can be considered as a suitable approximation of $h_{opt,0,k}$ as it is confirmed by the following theorem and its corollary.

Theorem 2.4. *Let the assumptions of Theorem 2.1 are fulfilled. Let $\mathcal{P}_0(h)$ stand for the left side of* (2.20) *and $\widehat{\mathcal{P}_0}(h)$ for the left side of* (2.24). *Then*

$$E(\widehat{\mathcal{P}_0}(h)) = \mathcal{P}_0(h) + o(h^{2k+1}),$$
$$\text{var}(\widehat{\mathcal{P}_0}(h)) = \frac{8k^2}{n^2 h}V(\Lambda)V(f) + o(n^{-2}h^{-1}). \qquad (2.26)$$

Theorem 2.4 states that $\widehat{\mathcal{P}_0(h)}$ is a consistent estimate of $\mathcal{P}_0(h)$. This result confirms that the solution of (2.25) may be expected to be reasonably close to h_{AMISE}.

Corollary 2.2. *The relative convergence rate of $\hat{h}_{\text{IT},0,k}$ to $h_{opt,0,k}$ (for case $k = 2$) can be expressed as*

$$\frac{\hat{h}_{\text{IT}} - h_{opt,0,2}}{h_{opt,0,2}} = O\left(n^{-\frac{1}{10}}\right).$$

See Complements for the proof.

Remark 2.6. Although the relative rate of convergence for bandwidths is the same for the cross-validation and the iterative method the constant in $O\left(n^{-\frac{1}{10}}\right)$ offers faster convergence for the iterative method. Moreover, the convolutions in the function Λ in (2.25) can be computed easily. Due to these facts, the iterative method seems to give better estimates of the optimal bandwidth and it is relatively fast. In paper Horová and Zelinka (2007a), it was shown that the iterative method is less time consuming than the cross-validation method.

2.5 Density derivative estimation

This paragraph is devoted to the kernel estimate of the ν-th derivative of the density f. We start with some additional assumptions:

- $\lim_{n\to\infty} h = 0$, $\lim_{n\to\infty} nh^{2\nu+1} = \infty$, $1 \le \nu$,
- $f \in C^{k_0}$, $\nu + k \le k_0$, $f^{(\nu+k)}$ is square integrable,
- $K \in S_{0,k}^\nu$, i.e., $K^{(\nu)} \in S_{\nu,k+\nu}^0$ (see Granovsky and Müller (1991); Marron and Nolan (1988)).

Usual kernel estimates for the ν-th derivative of f at a given point $x \in \mathbb{R}$ are defined by

$$\hat{f}^{(\nu)}(x,h) = \frac{1}{nh^{\nu+1}} \sum_{i=1}^{n} K^{(\nu)}\left(\frac{x - X_i}{h}\right). \qquad (2.27)$$

We again consider MISE as a criterion of the quality of the estimate

$$\mathrm{MISE}\{\hat{f}^{(\nu)}(\cdot,h)\} = E \int \{\hat{f}^{(\nu)}(x,h) - f^{(\nu)}(x)\}^2 dx.$$

The following theorem can be derived straightforwardly.

Theorem 2.5. *Under the above assumptions* $\mathrm{MISE}\{\hat{f}^{(\nu)}(\cdot,h)\}$ *takes the form*

$$\mathrm{MISE}\{\hat{f}^{(\nu)}(\cdot,h)\} = \frac{1}{nh^{2\nu+1}} V(K^{(\nu)}) + h^{2k}\frac{\beta_k^2(K)}{(k!)^2} V(f^{(k+\nu)})$$
$$+ o\left\{h^{2k} + (nh^{2\nu+1})^{-1}\right\},$$

i.e., $\hat{f}^{(\nu)}(\cdot,h)$ *is a consistent estimate of* $f^{(\nu)}$.

Using the properties of kernel K and its derivatives it is easy to show that

$$\beta_k(K) = \int_{-1}^{1} x^k K(x)dx = \frac{(-1)^\nu}{(k+1)\dots(k+\nu)} \int_{-1}^{1} x^{k+\nu} K^{(\nu)}(x)dx$$

and thus

$$\frac{\beta_k(K)}{k!} = (-1)^\nu \frac{\beta_{k+\nu}(K^{(\nu)})}{(k+\nu)!},$$

where $\beta_{k+\nu}(K^{(\nu)}) = \int\limits_{-1}^{1} x^{k+\nu} K^{(\nu)}(x)dx$.

Obviously, AMISE takes the form

$$\text{AMISE}\{\hat{f}^{(\nu)}(\cdot,h)\} = \frac{V(K^{(\nu)})}{nh^{2\nu+1}} + h^{2k}\beta_{k+\nu}^2(K^{(\nu)})D_{k+\nu} \qquad (2.28)$$

with

$$D_{k+\nu} = \frac{V(f^{(k+\nu)})}{((k+\nu)!)^2}$$

and, hence the optimal bandwidth minimizing AMISE

$$h_{opt,\nu,k+\nu}^{2(k+\nu)+1} = \frac{2\nu+1}{2knD_{k+\nu}}\gamma_{\nu,k+\nu}^{2(k+\nu)+1}, \qquad (2.29)$$

where

$$\gamma_{\nu,k+\nu}^{2(k+\nu)+1} = \frac{V(K^{(\nu)})}{\beta_{k+\nu}^2(K^{(\nu)})}.$$

It follows that AMISE-optimal bandwidth for estimating $f^{(\nu)}$ is of order $n^{-1/(2(k+\nu)+1)}$, compared to the optimal bandwidth of order $n^{-1/(2k+1)}$ for estimating f itself.

Evidently, we can again obtain the useful expression for $\text{AMISE}\{\hat{f}^{(\nu)}(\cdot,h_{opt,\nu,k+\nu})\}$

$$\text{AMISE}\{\hat{f}^{(\nu)}(\cdot,h_{opt,\nu,k+\nu})\} = T(K^{(\nu)})\frac{(2(k+\nu)+1)\gamma_{\nu,k+\nu}^{2\nu+1}}{2nkh_{opt,\nu,k+\nu}^{2\nu+1}}, \qquad (2.30)$$

where

$$T(K^{(\nu)}) = \left(|\beta_{k+\nu}(K^{(\nu)})|^{2\nu+1}V(K^{(\nu)})^k\right)^{\frac{2}{2(k+\nu)+1}}, \quad K^{(\nu)} \in S_{\nu,k+\nu}^0,$$

and

$$\text{AMISE}\{\hat{f}^{(\nu)}(\cdot,h_{opt,\nu,k+\nu})\} = O\left(n^{-\frac{2k}{2(k+\nu)+1}}\right).$$

It means that this rate becomes slower for higher values of ν reflecting the increasing difficulty inherent in the problems of estimating higher derivatives. Moreover it can be easy to show that the following lemma holds.

Lemma 2.5.

$$(2\nu + 1)\text{AIV}\{\hat{f}^{(\nu)}(\cdot, h_{opt,\nu,k+\nu})\} - 2k\text{AISB}\{\hat{f}^{(\nu)}(\cdot, h_{opt,\nu,k+\nu})\} = 0. \quad (2.31)$$

The approach discussed above will be very useful for an iteration method similar as for the density estimation.

Remark 2.7. We aim to estimate the optimal bandwidth for higher derivative by means of the bandwidth for density or its first derivative. The optimal bandwidth in (2.29) depends on $D_{k+\nu}$ but the optimal bandwidth for density itself depends only on D_k. To remedy this fact we modify a little our assumptions and assume that $K^{(\nu)} \in S^0_{\nu,k}$, i.e., $K \in S^\nu_{\nu,k-\nu}$, $f \in C^{k_0}$, $k \leq k_0$, $f^{(k)}$ is square integrable. It means that we use k instead of $k + \nu$ for $k \geq \nu + 2$, ν and k are of the same parity.

Then the analogous considerations as above lead to the following formula for AMISE:

$$\text{AMISE}\{\hat{f}^{(\nu)}(\cdot, h)\} = \frac{V(K^{(\nu)})}{nh^{2\nu+1}} + h^{2(k-\nu)}\beta_k^2(K^{(\nu)})D_k.$$

The optimal bandwidth takes the form

$$h_{opt,\nu,k}^{2k+1} = \frac{(2\nu + 1)}{2n(k - \nu)D_k}\gamma_{\nu,k}^{2k+1}, \quad (2.32)$$

where

$$\gamma_{\nu,k}^{2k+1} = \frac{V(K^{(\nu)})}{\beta_k^2(K^{(\nu)})}.$$

The formula (2.32) offers a very useful tool for the calculation the optimal bandwidth for $f^{(\nu)}$ by means of $h_{opt,0,k}$ and $h_{opt,1,k}$, respectively.

First, let ν and k be even integers. Then the use of (2.32) yields

$$h_{opt,\nu,k}^{2k+1} = \frac{(2\nu + 1)k}{k - \nu}\left(\frac{\gamma_{\nu,k}}{\gamma_{0,k}}\right)^{2k+1}h_{opt,0,k}^{2k+1}, \quad (2.33)$$

and if ν and k are odd integers

$$h_{opt,\nu,k}^{2k+1} = \frac{(2\nu + 1)(k - 1)}{3(k - \nu)}\left(\frac{\gamma_{\nu,k}}{\gamma_{1,k}}\right)^{2k+1}h_{opt,1,k}^{2k+1}. \quad (2.34)$$

Such a procedure is called a *factor method* (see, *e.g.*, Müller *et al.* (1987); Härdle *et al.* (2004); Horová *et al.* (2002)). The formulas above can be rewritten with $\hat{h}_{opt,\nu,k}$ instead of $h_{opt,\nu,k}$.

In this case Lemma 2.5 gives the relationship

$$2(k - \nu)\text{AISB}\{\hat{f}^{(\nu)}(\cdot, h_{opt,\nu,k})\} - (2\nu + 1)\text{AIV}\{\hat{f}^{(\nu)}(\cdot, h_{opt,\nu,k})\} = 0.$$

The $\text{AMISE}\{\hat{f}^{(\nu)}(\cdot, h)\}$ can be expressed by means of the functional T which now takes the form

$$T(K^{(\nu)}) = \left(V(K^{(\nu)})^{k-\nu}|\beta_k(K^{(\nu)})|^{2\nu+1}\right)^{\frac{2}{2k+1}}, \qquad (2.35)$$

and

$$\text{AMISE}\{\hat{f}^{(\nu)}(\cdot, h_{opt,\nu,k})\} = \frac{T(K^{(\nu)})\gamma_{\nu,k}^{2\nu+1}(2k + 1)}{2nh_{opt,\nu,k}^{2\nu+1}(k - \nu)}, \qquad (2.36)$$

which yields

$$\text{AMISE}\{\hat{f}^{(\nu)}(\cdot, h_{opt,\nu,k})\} = O\left(n^{-\frac{2(k-\nu)}{2k+1}}\right).$$

As earlier, the formula (2.36) is of a great interest since it will serve as a basis for constructing the automatic procedure for choosing the bandwidth, the kernel and its order simultaneously.

Remark 2.8. (Choosing the shape of the kernel)
We assume $K \in S_{\nu,k-\nu}^{\nu}$ and under the additional assumption that ν and k are of the same parity, $0 \le \nu \le k - 2$, we can proceed in the same way as in Sec. 2.3 and we use the optimal kernel minimizing functional (2.35).

2.5.1 *Choosing the bandwidth*

Most of popular bandwidth selection methods for density itself can be transferred into methods for density derivative estimators.

Cross-validation method, plug-in method, maximal smoothing principle

Härdle *et al.* (1990) proposed the modified cross-validation for the derivative estimate. The objective function is defined as

$$\text{CV}_{(\nu)}(h) = \int (\hat{f}^{(\nu)}(x, h))^2 dx - 2\frac{(-1)^\nu}{n} \sum_{i=1}^{n} \hat{f}_{-i}^{(2\nu)}(X_i, h),$$

where $\hat{f}^{(2\nu)}_{-i}$ is the estimate of the 2ν-th derivative of f at the point X_i without using this point.

The maximal smoothing principle, as well as the plug-in method, can be also modified for density derivative estimates.

Iterative method

As far as the iterative method (§2.4.5) is concerned we can again use the relation (2.31) and suitable estimate of the Asymptotic Integrated Variance and the Asymptotic Integrated Square Bias in the form

$$\widehat{\mathrm{AIV}}\{\hat{f}^{(\nu)}(\cdot, h)\} = \frac{1}{nh^{2\nu+1}} V(K^{(\nu)}),$$

$$\widehat{\mathrm{AISB}}\{\hat{f}^{(\nu)}(\cdot, h)\} = \int \left(\int K(x) \hat{f}^{(\nu)}(x - hy, h) dy - \hat{f}^{(\nu)}(x, h) \right)^2 dx$$

$$= \frac{(-1)^\nu}{n^2 h^{2\nu+1}} \sum_{\substack{i,j=1 \\ i \neq j}}^{n} \Lambda^{(2\nu)} \left(\frac{X_i - X_j}{h} \right),$$

where

$$\Lambda^{(2\nu)}(z) = \frac{d^{2\nu} \Lambda^{(0)}(z)}{dz^{2\nu}}$$

and

$$\Lambda^{(0)}(z) = \Lambda(z) = (K * K * K * K - 2K * K * K + K * K)(z)$$

(see also Definition 1.2).

In this case we consider the equation (as in 2.4.5)

$$\frac{2\nu + 1}{nh^{2\nu+1}} V(K^{(\nu)}) - 2k \frac{(-1)^\nu}{n^2 h^{2\nu+1}} \sum_{\substack{i,j=1 \\ i \neq j}}^{n} \Lambda^{(2\nu)} \left(\frac{X_i - X_j}{h} \right) = 0. \qquad (2.37)$$

Let $\hat{h}_{\mathrm{IT}, \nu, k}$ be a solution of this equation. This equation can be solved by a suitable numerical method. The statistical properties of this estimate and the rate of convergence have been developed and are given in the following theorem.

Theorem 2.6. *Let the assumptions of Theorem 2.5 be fulfilled and as above let $\mathcal{P}_\nu(h)$ stand for the left side of the equation (2.31) and $\widehat{\mathcal{P}_\nu}(h)$ for the left side of (2.37), respectively. Then*

$$\begin{aligned} E(\widehat{\mathcal{P}_\nu}(h)) &= \mathcal{P}_\nu(h) + o(h^{2k+1}) \\ \mathrm{var}(\widehat{\mathcal{P}_\nu}(h)) &= \frac{8k^2}{n^2 h^{4\nu+1}} V\left(\Lambda^{(2\nu)}\right) V(f) + o(n^{-2} h^{-(4\nu+1)}). \end{aligned} \qquad (2.38)$$

Proof.   The proof uses the result of Lemma 1.2 and can be found in Complements.   □

Corollary 2.3. *The relative convergence rate of* $\hat{h}_{IT,\nu,k}$ *to* $h_{opt,\nu,k}$ *can be expressed as*

$$\frac{\hat{h}_{IT,\nu,k} - h_{opt,\nu,k}}{h_{opt,\nu,k}} = O\left(n^{-\frac{1}{4(k+\nu)+2}}\right).$$

Proof.   The proof is shifted to Complements.   □

2.6   Automatic procedure for simultaneous choice of the kernel, the bandwidth and the kernel order

Returning to the asymptotic expressions of MISE in (2.12), (2.30) and (2.36) and taking into account the facts that we know the shape of the optimal kernel for each ν, k and that we know how to select an optimal bandwidth when k and K are fixed, it makes sense (see Horová *et al.* (2002)) to propose as a kernel order selection procedure the minimization with respect to k of

$$L(k) = T(K_{opt}^{(\nu)}) \frac{\gamma_{\nu,k}^{2\nu+1}(2k+1)}{2n(k-\nu)h_{opt,\nu,k}^{2\nu+1}}. \tag{2.39}$$

In view of the fact that ν and k should be of the same parity we only look for order k in the following set

$$I_\nu(k_0) = \left\{2j, \ j = 0, \ldots, \left[\frac{k_0 - \nu}{2}\right]\right\}.$$

Finally, we propose to use, as the selected kernel order, the value $\hat{k}$ such that

$$\hat{k} = \underset{k \in I_\nu(k_0)}{\arg\min} \hat{L}(k). \tag{2.40}$$

By combining the ideas developed in the previous paragraphs we arrive at the following algorithm that allows to choose all the parameters entering in the construction of kernel estimates. The main steps are now summarized. With respect to the fact that the majority of statisticians is interested in the estimation of f itself we present the special case when $\nu = 0$.

Description of the algorithm:

Step 1. For any $k \in I_0(k_0)$ find the optimal kernel $K_{opt} \in S_{0,k}^0$ given by
the formula in Theorem 1.2 or in Table 1.1 and compute the canonical
factor γ_{0k}.

Step 2. For any $k \in I_0(k_0)$ and kernel $K_{opt} \in S_{0,k}^0$ find the estimate
of the optimal bandwidth $\hat{h}_{opt,0,k} \in H_n$ ($H_n = [h_l, h_u]$, see §2.4.2)
using one of the methods given in Sec. 2.4.

Step 3. For any $k \in I_0(k_0)$ compute the selection criterion $L(k)$ in which
the values of K and $\hat{h}_{opt,0,k}$ are those obtained at Steps 1 and 2.

Step 4. Compute the optimal order $\hat{k}$ carrying out the minimization procedure (2.40) for $\nu = 0$.

Step 5. Use the parameters selected in the previous steps to get the optimal
kernel estimate of f, *i.e.*,

$$\hat{f}(x, \hat{h}_{opt,0,\hat{k}}) = \frac{1}{n\hat{h}_{opt,0,\hat{k}}} \sum_{i=1}^{n} K_{opt}\left(\frac{x - X_i}{\hat{h}_{opt,0,\hat{k}}}\right), \qquad K_{opt} \in S_{0,\hat{k}}^0.$$

$$(2.41)$$

The procedure for estimating $f^{(\nu)}$ can be realized in a similar way and
using the factor method for ν even. For instance, for $\nu = 2$ the formula
(2.33) gives

$$\hat{h}_{opt,2,k}^{2k+1} = \frac{5k}{k-2}\left(\frac{\gamma_{2,k}}{\gamma_{0,k}}\right)^{2k+1} \hat{h}_{opt,0,k}^{2k+1},$$

where $\gamma_{2,k}$ and $\gamma_{0,k}$ are obtained for the optimal $K^{(2)} \in S_{2,k}^0$ or $K \in S_{0,k-2}^2$,
respectively. In case ν is odd, it is necessary to estimate $\hat{h}_{opt,1,k}$ and use
the factor method (see Rem. 2.7 for details).

The estimate selected by the algorithm is shown to be asymptotically
optimal in terms of MISE among all possible choices, for the bandwidth, for
the kernel, and for its order. This fact was proven in Horová *et al.* (2002).

2.7 Boundary effects

Within this chapter, we have assumed that the density f satisfies certain
smoothness conditions over the whole real line. However there are many
densities that do not satisfy this condition. For example the exponential
density $f(x) = e^{-x}$, $x \geq 0$ has the discontinuity point $x = 0$ for the first
derivative. Figure 2.8 demonstrates problems of estimation of f near this
point. It shows a kernel estimate of f (solid curve) based on a sample size

$n = 100$ with the bandwidth $h = 0.786$. There is also the true density (dashed line) for comparison and the sample data (black and gray crosses). The black crosses represent observations in the "critical" boundary region $[0, h]$ and the gray crosses are for "inner" points (from (h, ∞)). The terminology for these problems is known as "boundary effects".

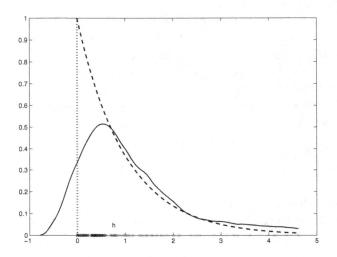

Fig. 2.8   Kernel density estimate near the boundary.

Now, we quantify these problems mathematically. Without the loss of generality, we will assume the support of f is $[0, \infty)$. Let us express a point x as $x = ch$, $c \geq 0$. As it was mentioned (see Sec. 2.2), for $x > h$ (*i.e.*, $c > 1$)

$$E\hat{f}(x, h) = f(x) + \frac{h^k \beta_k(K)}{k!} f^{(k)}(x) + o(h^k).$$

It means that for $c > 1$ $\hat{f}(x)$ is an asymptotically unbiased estimate of $f(x)$, moreover, Theorem 2.1 shows the consistency.

On the other hand, for $0 \leq c \leq 1$

$$E\hat{f}(x, h) = f(x) \int\limits_{-1}^{c} K(t)dt + o(1).$$

Since $\int\limits_{-1}^{c} K(t)dt \neq 1$ in general, the estimate $\hat{f}(x, h)$ is not consistent in such points.

To remove those boundary effects in kernel density estimation, a variety of methods have been developed in the literature. Some well-known methods are the reflection method (Cline and Hart (1991); Silverman (1986)); the boundary kernel method (Gasser *et al.* (1985); Jones (1993); Müller (1991); Zhang and Karunamuni (2000)); the transformation method (Marron and Ruppert (1994); Wand and Jones (1995)); the local linear method (Loader (1996); Hjort and Jones (1996); Cheng (1997); Zhang and Karunamuni (1998)) and some other methods (Zhang *et al.* (1999); Hall and Park (2002)).

In this section, we will follow the generalized reflection method in more details. The boundary kernel method (see Remark 1.4 for details) gives inappropriate results for density estimates as the estimate may be negative.

2.7.1 *Generalized reflection method*

This method is based on the reflection method which consists in reflecting data about zero and then estimating the density. It is defined by the formula

$$\hat{f}_R(x, h) = \frac{1}{nh} \sum_{i=1}^{n} \left\{ K\left(\frac{x - X_i}{h}\right) + K\left(\frac{x + X_i}{h}\right) \right\}. \qquad (2.42)$$

This makes $\hat{f}_R$ be a consistent estimate of f, but near the zero the bias is only of order $O(h)$ generally. For more see Schuster (1985); Silverman (1986); Cline and Hart (1991).

To improve the bias while holding onto the low variance, the generalized reflection method was proposed in Karunamuni and Alberts (2005a)

$$\hat{f}_G(x, h) = \frac{1}{nh} \sum_{i=1}^{n} \left\{ K\left(\frac{x - g_1(X_i)}{h}\right) + K\left(\frac{x + g_2(X_i)}{h}\right) \right\}, \qquad (2.43)$$

where g_1 and g_2 are some transformation functions. Generally, they are cubic polynomials with coefficients satisfying some specific criteria to make the bias of $\hat{f}_G$ of order $O(h^2)$. For more about the choice of g_1 and g_2 see Karunamuni and Alberts (2005b). Some statistical properties and improvements are in Karunamuni and Alberts (2006) and Karunamuni and Zhang (2008).

2.8 Simulations

The procedures for the bandwidth selection described in the previous sections were applied to simulated data. The samples contained simulated data from Example 2.1. In this case the optimal bandwidth can be evaluated as $h_{opt,0,2} = 0.4459$ for the Epanechnikov kernel.

Two hundred random samples were generated. For each data set, we estimated the optimal bandwidth by above-mentioned methods, *i.e.*, for each method we obtained 200 estimates. We compared the means and standard deviations of the estimates.

The means $E(\hat{h})$ and the standard deviations $std(\hat{h})$ of all estimates of the optimal bandwidth are presented in Table 2.2. It is obvious that the iterative method yields very good result in terms of mean. The best method with respect to $E(\hat{h})$ is the least squares cross-validation method but its standard deviation is not so good as in case of the iterative method.

Table 2.2   Estimates of the optimal bandwidth for simulated data

method	$E(\hat{h})$	$std(\hat{h})$
reference rule	0.9525	0.0816
maximal smoothing	1.0287	0.0881
least squares cross-validation	0.5229	0.1544
biased cross-validation	0.7300	0.3057
plug-in	1.0266	0.0853
iterative	0.5809	0.0775

2.9 Application to real data

2.9.1 *Buffalo snowfall data*

Buffalo snowfall (see Scott (1992)) data set is interesting because it is not easy to decide between unimodality and trimodality of the estimate of the density function (see also Minnotte (1993) for details). The data set contains measurements of snowfall in Buffalo, New York, from 1910 to 1972. The authors of various papers are not of the same opinion in the interpretation of data. Figure 2.9 presents two estimates of the density function.

The trimodal estimate (dashed line) is obtained for the bandwidth given by the least squares cross-validation method, the unimodal estimate (solid line) by the bandwidth given by the iterative method.

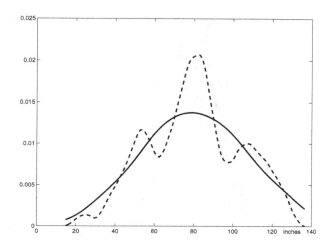

Fig. 2.9   Densities for Buffalo snowfall data.

Table 2.3 brings the estimates of the optimal bandwidth.

Table 2.3   Estimates of the optimal bandwidth for Buffalo snowfall data

method	estimate
reference rule	24.2868
maximal smoothing	26.2282
least squares cross-validation	16.8148
biased cross-validation	21.5577
plug-in	32.7051
iterative	29.7976

2.9.2   *Concentration of cholesterol*

The last data set comes also from Scott (1992). It contains the concentration of cholesterol for 371 patients. The biased cross-validation bandwidth leads to undersmoothed estimate (dashed line) and the plug-in bandwidth $\hat{h}_{\mathrm{PI}}$ yields much worse estimate, the iterative bandwidth (solid line) gives much more better estimate (see Fig. 2.10).

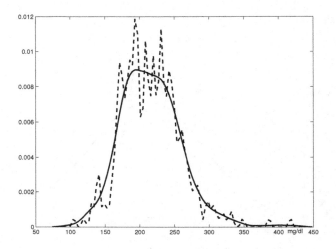

Fig. 2.10   Concentration of cholesterol.

The estimates of the optimal bandwidth can be found in Table 2.4.

Table 2.4   Estimates of the optimal bandwidth for cholesterol data

method	estimate
reference rule	31.8209
maximal smoothing	33.7571
least squares cross-validation	28.4487
biased cross-validation	16.0353
plug-in	42.8912
iterative	27.8974

Our experience show that the presented iterative method seems to be a suitable tool for the choice of the bandwidth. It seems to be sufficiently reliable and less time consuming than, *e.g.*, cross-validation method.

2.10 Use of MATLAB toolbox

The toolbox can be downloaded from the web page
http://www.math.muni.cz/english/science-and-research/
developed-software/232-matlab-toolbox.html.

2.10.1 *Running the program*

Toolbox for kernel density estimates can be launched by command **ksdens**. Launching without parameters will cause the start to the situation when only data input (button $\textcircled{1}$) or terminating the program (button $\textcircled{2}$) is possible (see Fig. 2.11). In the data input subroutine (Fig. 2.12) you can

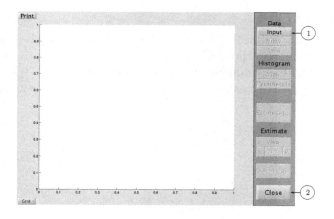

Fig. 2.11   Start of the program.

choose reading data from the workspace (button $\textcircled{3}$), from the external file (button $\textcircled{4}$) or to create simulated data (button $\textcircled{5}$). After choosing the source of data select the name of the variable containing the random variable (button $\textcircled{6}$). If you know the true density (*e.g.*, for simulated data), you can put it to the text field $\textcircled{7}$ with 'x' as variable. It can be used to compare with the final estimate of the density.

At the end you can cancel the subroutine (button $\textcircled{8}$) or confirm data (button $\textcircled{9}$).

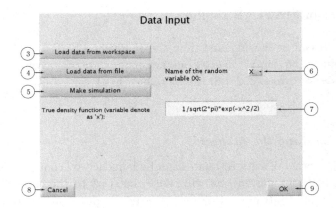

Fig. 2.12   Data input.

2.10.2   *Main figure*

After data input you will see data and you obtain another possibilities (Fig. 2.13) in the main figure. The same situation is invoked if the main program is called with a parameter, *i.e.*, by command **ksdens(X)**. The variable **X** contains a sample of random variable which density we want to estimate.

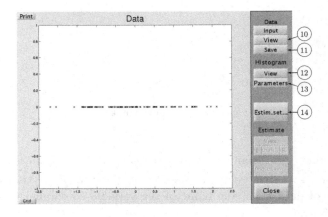

Fig. 2.13   Data view.

Pressing button ⑩ calls the same figure as Fig. 2.13. It is also possible to save selected variables (button ⑪, see Fig. 2.14) into a file (button ⑮)

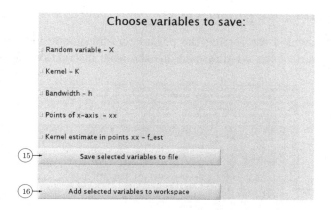

Fig. 2.14   Data saving.

or into the workspace (button ⑯). It should be point out that the most of variables is empty at the beginning of the program.

Another two possibilities after data input concern the histogram: you can display it by button ⑫ (see Fig. 2.15) and set the number of bins for the histogram by button ⑬.

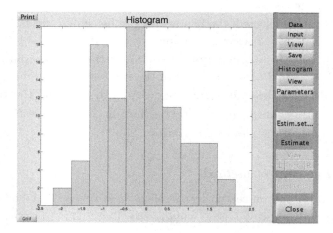

Fig. 2.15   Histogram.

2.10.3 *Setting the parameters*

Button ⑭ invokes the setting the parameters for smoothing (Fig. 2.16). The upper part of the window contains the brief instructions for entering

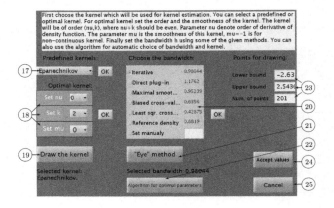

Fig. 2.16   Setting parameters.

the parameters. In the left part of the window you can choose a predefined kernel (button ⑰) or the optimal kernel (buttons ⑱). Confirm the choice of the kernel by pressing button OK. By button ⑲ you get the picture with the shape of the kernel.

If the kernel is selected, the bandwidth can be chosen (field ⑳). Basic methods of bandwidth selection are implemented (see Sec. 2.4 for details). Button ㉒ calls the automatic procedure for setting of all parameters described in Sec. 2.6. In the right part of the window there are boxes where points for drawing the estimate of the density can be set up (button ㉓). Finally, you can confirm the parameters (button ㉔) or close the window without change of the parameters (button ㉕).

2.10.4 *Eye-control method*

The bandwidth can be chosen by so-called *"Eye-control" method* (button ㉑). This button invokes other window (see Fig. 2.17). In boxes ㉖ and ㉗ the bandwidth and the step for its increasing or decreasing can be set. By pressing button ㉘ (the middle one) the estimate for actual bandwidth is displayed, the arrows at the left and right side cause increasing or de-

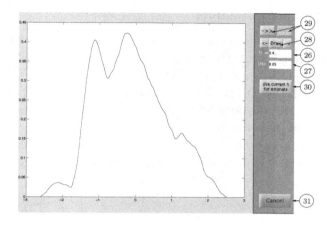

Fig. 2.17   "Eye-control" method.

creasing the bandwidth by a step and redrawing the figure. By buttons
(29) you can run and stop the gradual increasing the bandwidth and drawing the corresponding estimate. Finally it is possible to accept selected
bandwidth (button (30)) or cancel the procedure (button (31)).

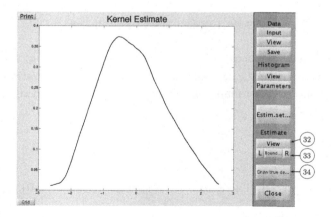

Fig. 2.18   Kernel estimate.

2.10.5   *The final estimation*

If all smoothing parameters are set, the estimate of the density function can be displayed (Fig. 2.18, button ㉜). For correction of boundary effects buttons ㉝ can be applied. In the separate window you can set the left and the right boundaries for the removal the boundary effects. Then pressing buttons "L" or "R" the boundary effects correction is applied at the corresponding part of the estimate.

Button ㉞ is intended to display the true density function if it was specified (see Fig. 2.19). Button ㉟ can be used for displaying the grid

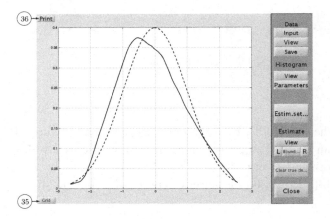

Fig. 2.19   Estimate with the true density.

in the figure that may help to identify the values of the graph. By the last button ㊱ you obtain the separate window with the density estimate without the buttons for other manipulation with the graph (exporting into some graphical format *etc.*)

2.11   Complements

Proof of Theorem 2.6 (Theorem 2.4 is a special case for $\nu = 0$).

Proof.   Let us denote

$$\mathcal{P}_\nu(h) = \frac{2\nu + 1}{nh^{2\nu+1}} V(K^{(\nu)}) - 2kh^{2k} \left(\frac{\beta_{k+\nu}}{(k+\nu)!} \right)^2 V(f^{(\nu+k)}) \qquad (2.44)$$

and let

$$\widehat{\mathcal{P}_\nu}(h) = \frac{2\nu + 1}{nh^{2\nu+1}} V(K^{(\nu)}) - \frac{2k(-1)^\nu}{n^2} \sum_{\substack{i,j=1 \\ i \neq j}}^{n} \Lambda_h^{(2\nu)}(X_i - X_j) \qquad (2.45)$$

stand for an estimate of $\mathcal{P}_\nu$. The proposed method aims to solve the equation

$$\widehat{\mathcal{P}_\nu}(h) = 0.$$

We start with an evaluation of $E\Lambda_h^{(2\nu)}(X_1 - X_2)$:

$$\begin{aligned}
E\Lambda_h^{(2\nu)}(X_1 - X_2) &= \frac{1}{h^{2\nu+1}} \int \int \Lambda^{(2\nu)}\left(\frac{x-y}{h}\right) f(x)f(y)dxdy \\
&= \int \int \Lambda(u) f^{(2\nu)}(x - uh) f(x) dx du \\
&= \int \int f(x)\Lambda(u)\{f^{(2\nu)}(x) + \dots \\
&\quad + \frac{(-1)^{2k}}{(2k)!} f^{2(\nu+k)}(x) h^{2k} u^{2k} + o(h^{2k})\} dx du.
\end{aligned}$$

Using Lemma 1.2 leads to the result

$$E\Lambda_h^{(2\nu)}(X_1 - X_2) = (-1)^{\nu+k} h^{2k}\left(\frac{\beta_{k+\nu}}{(k+\nu)!}\right)^2 V(f^{(\nu+k)}) + o(h^{2k}).$$

Thus

$$\begin{aligned}
E\widehat{\mathcal{P}_\nu}(h) &= \frac{2\nu+1}{nh^{2\nu+1}} V(K^{(\nu)}) - \\
&\quad - 2k(1 - n^{-1})h^{2k}\left(\frac{\beta_{k+\nu}}{(k+\nu)!}\right)^2 V(f^{(\nu+k)}) + o(h^{2k})
\end{aligned}$$

and

$$E\widehat{\mathcal{P}_\nu}(h) = \mathcal{P}_\nu(h) + o(h^{2k}). \qquad (2.46)$$

Since it is assumed $\lim_{n\to\infty} nh^{2\nu+1} = \infty$ then $E\widehat{\mathcal{P}_\nu}(h) \to \mathcal{P}_\nu(h)$.

Now, we derive the formula for $\mathrm{var}\widehat{\mathcal{P}_\nu}(h)$. Our considerations are based on the well-known statement (see, *e.g.*, Wand and Jones (1995)):
Let $X_1, \dots, X_n$ be i.i.d. random variables and define

$$U = 2n^{-2} \sum_{i=1}^{n-1} \sum_{j=i+1}^{n} S(X_i - X_j),$$

where S is a symmetric function about zero.
Then

$$
\begin{aligned}
\mathrm{var}U = {} & 2n^{-3}(n-1)\mathrm{var}S(X_1 - X_2) \\
& + 4n^{-3}(n-1)(n-2)cov\{S(X_1 - X_2), S(X_2 - X_3)\}.
\end{aligned}
\tag{2.47}
$$

Let

$$
I_1 = E\left(\Lambda_h^{(2\nu)}(X_1 - X_2)\right)^2
$$

$$
I_2 = E\left[\Lambda_h^{(2\nu)}(X_1 - X_2)\Lambda_h^{(2\nu)}(X_2 - X_3)\right].
$$

As concerns the quantity I_1 it is easy to show that

$$
\begin{aligned}
I_1 &= \int\int \frac{1}{h^{4\nu+2}}\left(\Lambda^{(2\nu)}\left(\frac{x-y}{h}\right)\right)^2 f(x)f(y)dxdy \\
&= \frac{1}{h^{4\nu+1}}\int\int\left(\Lambda^{(2\nu)}(u)\right)^2 f(x-uh)f(x)dudx \\
&= \frac{1}{h^{4\nu+1}}\int\int\left(\Lambda^{(2\nu)}(u)\right)^2\{f(x)+o(1)\}f(x)dudx \\
&= \frac{1}{h^{4\nu+1}}V\left(\Lambda^{(2\nu)}\right)V(f) + o(h^{-(4\nu+1)})
\end{aligned}
$$

and

$$
\begin{aligned}
I_2 &= \int\int\int \Lambda_h^{(2\nu)}(x-y)\Lambda_h^{(2\nu)}(y-z)f(x)f(y)f(z)dxdydz \\
&= \int\int\int \Lambda(u)\Lambda(v)f(y)f^{(2\nu)}(y+uh)f^{(2\nu)}(y-vh)dudvdy \\
&= \int f(y)\int\Lambda(u)\left\{f^{(2\nu)}(y) + \cdots + \frac{(uh)^{2k}}{(2k)!}f^{(2(\nu+k))}(y) + o(h^{2k})\right\}du \\
&\quad \times \int\Lambda(v)\left\{f^{(2\nu)}(y) + \cdots + \frac{(vh)^{2k}}{(2k)!}f^{(2(\nu+k))}(y) + o(h^{2k})\right\}dvdy.
\end{aligned}
$$

If Lemma 1.2 is again used, we arrive at the following expression of I_2

$$
I_2 = h^{4k}\left(\frac{\beta_{k+\nu}}{(k+\nu)!}\right)^4\int f(y)\left(f^{(2(\nu+k))}(y)\right)^2 dy + o(h^{4k}).
$$

Combining I_1 and I_2 with the formula (2.47) leads to

$$
\mathrm{var}\widehat{\mathcal{P}_\nu}(h) = \frac{8k^2}{n^2 h^{4\nu+1}}V\left(\Lambda^{(2\nu)}\right)V(f) + o(n^{-2}h^{-(4\nu+1)}).
\tag{2.48}
$$

Taking (2.46) and (2.48) into account Theorem 2.4 and Theorem 2.6 follow.

$\square$

Sketch of the proof of Corollary 2.3:

Proof. Using notation from the previous proof we can expand

$$0 = (\widehat{\mathcal{P}_\nu} - \mathcal{P}_\nu)(\hat{h}_{\mathrm{IT},\nu,k}) + \mathcal{P}_\nu(\hat{h}_{\mathrm{IT},\nu,k})$$
$$= (1 + o(1))(\widehat{\mathcal{P}_\nu} - \mathcal{P}_\nu)(h_{opt,\nu,k}) + \mathcal{P}_\nu(h_{opt,\nu,k})$$
$$+ (1 + o(1))\mathcal{P}'_\nu(h_{opt,\nu,k})(\hat{h}_{\mathrm{IT},\nu,k} - h_{opt,\nu,k}).$$

After removing all negligible terms we obtain

$$0 = (\widehat{\mathcal{P}_\nu} - \mathcal{P}_\nu)(h_{opt,\nu,k}) + \mathcal{P}'_\nu(h_{opt,\nu,k})(\hat{h}_{\mathrm{IT},\nu,k} - h_{opt,\nu,k}).$$

It is easy to see that $\mathcal{P}'_\nu(h_{opt,\nu,k}) = Cn^{-(2k-1)/(2k+2\nu+1)}$ for a constant C, which implies

$$\hat{h}_{\mathrm{IT},\nu,k} - h_{opt,\nu,k} = -C^{-1}n^{(2k-1)/(2k+2\nu+1)}(\widehat{\mathcal{P}_\nu} - \mathcal{P}_\nu)(h_{opt,\nu,k}).$$

Since $\mathrm{MSE}\left\{(\widehat{\mathcal{P}_\nu} - \mathcal{P}_\nu)(h_{opt,\nu,k})\right\} = O(n^{-(4k+1)/(2k+2\nu+1)})$, then

$$\mathrm{MSE}\{\hat{h}_{\mathrm{IT},\nu,k} - h_{opt,\nu,k}\} = C^{-2}n^{(4k-2)/(2k+2\nu+1)}\mathrm{MSE}\left\{(\widehat{\mathcal{P}_\nu} - \mathcal{P}_\nu)(h_{opt,\nu,k})\right\}$$
$$= O\left(n^{-1/(2k+2\nu+1)}\right)h^2_{opt,\nu,k}$$

and hence,

$$\frac{\hat{h}_{\mathrm{IT},\nu,k} - h_{opt,\nu,k}}{h_{opt,\nu,k}} = O\left(n^{-\frac{1}{4(k+\nu)+2}}\right).$$

$\square$

Chapter 3

Kernel estimation of a distribution function

The most commonly used nonparametric estimation of a distribution function F is an empirical distribution function F_n. But F_n is a step function even in case that F is continuous. Nadaraya (1964) proposed a smooth non-parametric alternative to F_n, namely, kernel distribution estimator. This estimator $\widehat{F}$ is obtained by integrating the well known Rosenblatt-Parzen kernel density estimate (2.2). More generally, having a nonparametric estimate $\hat{f}$ of f one can obtain a nonparametric estimate of F by integrating $\hat{f}$.

Nadaraya (1964) proved that $\widehat{F}$ is asymptotically unbiased and has the same variance as F_n. Moreover, he obtained uniform convergence to F with probability one. Watson and Leadbetter (1964) derived the asymptotic normality. Reiss (1981) derived an asymptotic expression for bias and variance of the estimate. When applying $\hat{f}$ one needs to choose the kernel and the smoothing parameter. It is shown in Lejeune and Sarda (1992), from an empirical study that the choice of the kernel is less important than the choice of the smoothing parameter for the performance of the estimate, a fact recognized in kernel density and thus the choice of a smoothing parameter will be only treated.

3.1 Basic definition

Let $X_1, \ldots, X_n$ be independent real random variables with a common distribution function F and a density f. The natural estimate of F, from $X_1, \ldots, X_n$, is the so-called empirical distribution function F_n defined at some point x as

$$\widehat{F}_n(x) = \frac{1}{n} \sum_{i=1}^{n} I_{(-\infty, x]}(X_i).$$

Despite of the good statistical properties of $\widehat{F}_n$ one could prefer a rather smooth estimate. This fact leads to the effort to obtain a smooth estimate (see Azzalini (1981); Lejeune and Sarda (1992); Bowman *et al.* (1998); Altman and Léger (1995); Sarda (1993); Horová *et al.* (2008a)).

Nadaraya (1964) obtained a smooth estimate of F by integrating a kernel estimate of density:

$$\widehat{F}(x,h) = \int\limits_{-\infty}^{x} \widehat{f}(t,h)dt = \frac{1}{nh}\sum_{i=1}^{n}\int\limits_{-\infty}^{x} K\left(\frac{t-X_i}{h}\right)dt.$$

Using the substitution $y = (t - X_i)/h$ leads to

$$\widehat{F}(x,h) = \frac{1}{n}\sum_{i=1}^{n}\int\limits_{-\infty}^{\frac{x-X_i}{h}} K(y)dy = \frac{1}{n}\sum_{i=1}^{n} W\left(\frac{x-X_i}{h}\right)$$

as K is supported on $[-1,1]$.

Then the kernel estimate of F at the point $x \in \mathbb{R}$ is defined as

$$\widehat{F}(x,h) = \frac{1}{n}\sum_{i=1}^{n} W\left(\frac{x-X_i}{h}\right), \ W(x) = \int\limits_{-1}^{x} K(t)dt, \qquad (3.1)$$

where K, $K \geq 0$, is assumed to belong to the class $S_{0,2}$.

Example 3.1. See Fig. 3.1 and Fig. 3.2 for comparing empirical and kernel estimates of a distribution function. Figures illustrate the estimations based on 30 observations of the standardized normal distribution. The dashed line represents the original distribution function, the full line represents the kernel estimate.

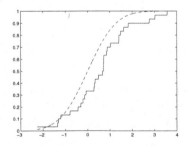

Fig. 3.1   Empirical estimate.

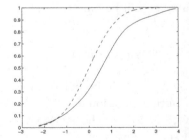

Fig. 3.2   Kernel estimate with $h = 1$.

Below, the properties and statements of the function W are summarized:

1° $W(x) = 0$ for $\forall x \in (-\infty, -1]$, $W(x) = 1$ for $\forall x \in [1, \infty)$.

2° $\int\limits_{-1}^{1} W^2(x)dx \leq \int\limits_{-1}^{1} W(x)dx = 1$.

3° $\int\limits_{-1}^{1} W(x)K(x)dx = \frac{1}{2}$.

4° $\int\limits_{-1}^{1} xW(x)K(x)dx = \frac{1}{2}\left(1 - \int\limits_{-1}^{1} W^2(x)dx\right)$.

Example 3.2. If we use the Epanechnikov kernel $K(x) = \frac{3}{4}(1 - x^2)I_{[-1,1]}(x)$, then the function $W(x)$ takes the form

$$W(x) = \begin{cases} 0, & x \leq -1, \\ -\frac{1}{4}x^3 + \frac{3}{4}x + \frac{1}{2}, & |x| \leq 1, \\ 1, & x \geq 1. \end{cases}$$

See Fig. 3.3 and Fig. 3.4.

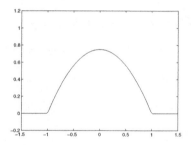

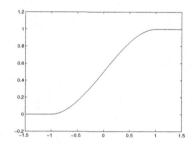

Fig. 3.3 $K(x) = \frac{3}{4}(1 - x^2)I_{[-1,1]}(x)$, the Epanechnikov kernel.

Fig. 3.4 $W(x) = \int\limits_{-\infty}^{x} \frac{3}{4}(1 - t^2) \times I_{[-1,1]}(t)dt.$

3.2 Statistical properties of the estimate

A useful criterion for the quality when estimating at a single point is the Mean Square Error (MSE) defined by

$$\text{MSE}\{\widehat{F}(x, h)\} = E\{\widehat{F}(x, h) - F(x)\}^2.$$

In Complements, it is shown that $\mathrm{MSE}\{\widehat{F}(x,h)\}$ (provided that $F \in C^2$, $h = h(n)$, $h(n) \to 0$ and $nh(n) \to \infty$ as n tends to infinity) takes the form

$$\mathrm{MSE}\{\widehat{F}(x,h)\} = \frac{1}{n}F(x)(1 - F(x)) - \frac{1}{n}hf(x)\left(1 - \int_{-1}^{1}W^2(t)dt\right)$$
$$+ \frac{1}{4}(F''(x))^2h^4\beta_2^2 + o\left(\frac{h}{n} + h^4\right). \tag{3.2}$$

Two global measures of quality of the kernel estimator are the Integrated Square Error

$$\mathrm{ISE}\{\widehat{F}(\cdot,h)\} = \int \{\widehat{F}(x,h) - F(x)\}^2 dx$$

and the Mean Integrated Square Error

$$\mathrm{MISE}\{\widehat{F}(\cdot,h)\} = E \int \{\widehat{F}(x,h) - F(x)\}^2 dx.$$

Remark 3.1. It can be shown (see Complements for a proof) by means of integration by parts that

$$E\widehat{F}(x,h) - F(x) = \int_{-1}^{1} K(t)F(x - ht)dt - F(x)$$

and thus

$$\mathrm{ISB}\{\widehat{F}(\cdot,h)\} = \int \left(\int_{-1}^{1} K(t)F(x - ht)dt - F(x)\right)^2 dx$$
$$= \int \left((K_h * F)(x) - F(x)\right)^2 dx.$$

We proceed in a similar way as in the kernel density estimate and investigate approximation of $\mathrm{MISE}\{\widehat{F}(\cdot,h)\}$. The following theorem yields such an expression.

Theorem 3.1. *Let $F \in C^2$, F'' be square integrable, the kernel K belong to the class $S_{0,2}$ and $\lim_{n\to\infty} h = 0$, $\lim_{n\to\infty} nh = \infty$. Then*

$$\mathrm{MISE}\{\widehat{F}(\cdot,h)\} = \frac{1}{n}\int F(x)(1 - F(x))dx - c_1\frac{h}{n} + c_2h^4$$
$$+ o\left(\frac{h}{n} + h^4\right), \tag{3.3}$$

where

$$c_1 = 1 - \int_{-1}^{1} W^2(t)dt, \quad c_2 = \frac{\beta_2^2}{4} \int \left(F''(t)\right)^2 dt.$$

Proof. See Complements for a proof. □

Corollary 3.1. *Under the given assumptions $\widehat{F}(\cdot, h)$ is a consistent estimate of F.*

From the same reason as in kernel density estimates we employ the Asymptotic Mean Integrated Square Error (AMISE) as a sum of the Asymptotic Integrated Variance and the Asymptotic Integrated Square Bias

$$\text{AMISE}\{\widehat{F}(\cdot, h)\} = \underbrace{\frac{1}{n} \int F(x)(1 - F(x))dx - c_1\frac{h}{n}}_{\text{AIV}} + \underbrace{c_2 h^4}_{\text{AISB}}. \qquad (3.4)$$

Let h_{opt}^F stand for a bandwidth minimizing $\text{AMISE}\{\widehat{F}(\cdot, h)\}$, *i.e.*,

$$h_{opt}^F = \arg\min \text{AMISE}\{\widehat{F}(\cdot, h)\}.$$

Then it is easy to show that

$$h_{opt}^F = n^{-1/3} \left(\frac{c_1}{4c_2}\right)^{1/3}, \text{ i.e., } h_{opt}^F = O\left(n^{-1/3}\right) \qquad (3.5)$$

and as a consequence

$$\text{AMISE}\{\widehat{F}(\cdot, h_{opt}^F)\} = \frac{1}{n} \int F(x)(1 - F(x))dx - \frac{3}{c_2^{1/3}} \left(\frac{c_1}{4}\right)^{4/3} n^{-4/3}. \qquad (3.6)$$

Here, we emphasize that for the kernel density estimate with $K \in S_{0,2}$ the optimal bandwidth is of order $n^{-1/5}$.

Lemma 2.4 shows an interesting relation between $\text{AIV}\{\hat{f}(\cdot, h_{opt})\}$ and $\text{AISB}\{\hat{f}(\cdot, h_{opt})\}$. We aim to find a similar relation in the kernel estimation of the distribution function.

We remove the nonstochastic part $\frac{1}{n} \int F(x)(1 - F(x))dx$ from the $\text{AMISE}\{\widehat{F}(\cdot, h)\}$ formula and consider

$$\text{AMISE}^*\{\widehat{F}(\cdot, h)\} = -c_1\frac{h}{n} + c_2 h^4.$$

Now, we define

$$\text{AIV}^*\{\widehat{F}(\cdot, h)\} = \text{AIV}\{\widehat{F}(\cdot, h)\} - \frac{1}{n} \int F(x)(1 - F(x))dx.$$

It is easy to show that for h_{opt}^F the following lemma holds.

Lemma 3.1.

$$-\text{AIV}^*\{\widehat{F}(\cdot, h_{opt}^F)\} = 4\text{AISB}\{\widehat{F}(\cdot, h_{opt}^F)\}, \qquad (3.7)$$

or

$$\frac{c_1 h_{opt}^F}{n} = 4c_2 (h_{opt}^F)^4.$$

This lemma again shows the balance between the asymptotic integrated variance and the asymptotic integrated square bias.

Remark 3.2. (*Choosing the shape of the kernel*)
As it was mentioned, the choice of the kernel is less important than the bandwidth choice. We recommend to use smooth kernels of the class $S_{0,2}^\mu$, $\mu = 0, 1, 2$, *e.g.*, the Epanechnikov, quartic, triweight kernel.

3.3 Choosing the bandwidth

3.3.1 *Cross-validation methods*

The well-known methods are methods based on cross-validation ideas (see Sarda (1993); Altman and Léger (1995); Bowman *et al.* (1998)). Here, we only recall the method proposed by Bowman *et al.* (1998). They considered a cross-validation function of the form

$$\text{CV}_\text{B}(h) = \frac{1}{n} \sum_{i=1}^n \int \left(I_{(-\infty, x]}(X_i) - \widehat{F}_{-i}(x, h) \right)^2 dx,$$

where $\widehat{F}_{-i}(x, h)$ is a kernel estimate based on the sample with X_i deleted. Let

$$\hat{h}_{\text{CV}_\text{B}}^F = \arg \min_{h \in H_n} \text{CV}_\text{B}(h)$$

where commonly $H_n = [\alpha n^{-1/3}, \beta n^{-1/3}]$ for some $0 < \alpha < \beta < \infty$ (see more in the next paragraph).

Under some additional assumptions it can be shown that

$$\frac{\hat{h}_{\text{CV}_\text{B}}^F - h_{opt}^F}{h_{opt}^F} = O\left(n^{-\frac{1}{6}}\right),$$

see Bowman *et al.* (1998).

3.3.2 *Maximal smoothing principle*

We use the fact that

$$\int F''^2(x)dx = \int f'^2(x)dx$$

we can apply Theorem 2.3 with $k = 1$. In this case,

$$g_1(x) = \frac{15}{16}(1 - x^2)^2, \ |x| \leq 1$$

and then

$$h^F_{opt} = n^{-1/3} \left(\frac{c_1}{\beta_2^2 \int f'^2(x)dx} \right)^{1/3}$$

$$\leq n^{-1/3} \left(\frac{c_1}{\beta_2^2} \right)^{1/3} \frac{\sigma}{\sigma_1} \left(\int\limits_{-1}^{1} g_1'^2(x)dx \right)^{-1/3},$$

where $\sigma_1 = \int\limits_{-1}^{1} x^2 g_1(x)dx = \frac{1}{7}$, $\int\limits_{-1}^{1} g_1'^2(x)dx = \frac{15}{7}$.

Thus

$$\hat{h}^F_{MS} = n^{-1/3} \left(\frac{7c_1}{15\beta_2^2} \right)^{1/3} \sqrt{7}\hat{\sigma}, \tag{3.8}$$

$\hat{\sigma}$ is an estimate of σ (see equations (2.13) and (2.14)).

The value h^F_{MS} can serve as an upper bound for the set of smoothing parameters for CV methods and $H_n = [h_l, h^F_{MS}]$, h_l is a minimum distance between consecutive points X_i, $i = 1, \ldots, n$.

3.3.3 *Plug-in methods*

These methods aim to estimate $\int F''^2(x)dx$ and we used notation $\hat{h}_{PI}$ for such bandwidths. Under sufficient smoothness assumptions on f one obtains using integration by parts

$$\int F''^2(x)dx = -\int f''(x)f(x)dx.$$

It is, therefore, purposeful to study an estimation of

$$\psi_1 = \int f''(x)f(x)dx.$$

Note that

$$\psi_1 = E\{f''(X)\}. \tag{3.9}$$

This motivates the estimator (see §2.4.4)

$$\widehat{\psi}_1(g) = n^{-1} \sum_{i=1}^{n} \hat{f}''(X_i, g) = n^{-2} \sum_{i=1}^{n} \sum_{j=1}^{n} L_g''(X_i - X_j) \qquad (3.10)$$

(Hall and Marron (1987); Jones and Sheather (1991)), where g and L are a bandwidth and a kernel, respectively, that can be different form h and K.

Altman and Léger (1995) proposed

$$\widehat{\psi}_1(h) = n^{-2} h^{-3} \sum_{i=1}^{n} \sum_{j=1}^{n} L'' \left(\frac{X_i - X_j}{h} \right) \qquad (3.11)$$

with $L(x) = \frac{1}{\sqrt{2\pi}} e^{-x^2/2}$ and they put $h = n^{-0.3}$.

As an alternative to this approach we proposed a method based on the kernel estimate of the second derivative. The formula (2.33) is used for estimating the second derivative with the kernel $K^{(2)} \in S_{2,4}^0$. Thus

$$\widehat{\psi}_1 = n^{-1} \sum_{i=1}^{n} \hat{f}''(X_i, h) = n^{-2} h^{-3} \sum_{i=1}^{n} \sum_{j=1}^{n} K^{(2)} \left(\frac{X_i - X_j}{h} \right), \qquad (3.12)$$

where $h = h_{opt,2,4}$ and hence

$$\widehat{c}_2 = -\frac{\beta_2^2}{4} \widehat{\psi}_1.$$

Remark 3.3. Especially, the estimate of optimal bandwidth for the Epanechnikov kernel is

$$\hat{h}_{PI}^F = n^{-1/3} \left(\frac{45}{-7\widehat{\psi}_1} \right)^{1/3}$$

and for the quartic kernel

$$\hat{h}_{PI}^F = n^{-1/3} \left(\frac{350}{-33\widehat{\psi}_1} \right)^{1/3}.$$

Summarizing our considerations and calculations we arrive at the procedure of estimating F.

Description of the algorithm:

Step 1. Find the estimate of the optimal bandwidth $h_{opt,0,4}$ for a density with the optimal kernel $K_{opt} \in S_{0,4}^0$;

Step 2. Find the estimate of the optimal bandwidth $h_{opt,2,4}$ for the second derivative by the formula (2.33) with $k = 4$ (and the optimal kernel $K_{opt}^{(2)} \in S_{2,4}^0$);

Step 3. Compute the estimate $\widehat{\psi}_1$ using the kernel $K_{opt}^{(2)}$ and the bandwidth in the preceding step;

Step 4. Evaluate

$$\hat{h}_{\text{PI}}^F = n^{-1/3} \left(\frac{c_1}{-\widehat{\psi}_1 \beta_2^2} \right)^{1/3};$$

Step 5. Find the estimate $\widehat{F}(x, h)$ with the given kernel $K \in S_{0,2}^\mu$, $\mu = 0, 1, 2$ (the Epanechnikov kernel or the quartic kernel are recommended).

3.3.4 *Iterative method*

Like in density estimates a method based on equation (3.7) can be developed. A suitable estimate of AISB$\{\widehat{F}(\cdot, h)\}$ is considered as

$$\overline{\text{AISB}}\{\widehat{F}(\cdot, h)\} = \int \left((K_h * \widehat{F})(x, h) - \widehat{F}(x, h) \right)^2 dx$$

$$= -\frac{h}{n^2} \sum_{i,j=1}^n (K * K * W * W - 2K * W * W$$

$$+ W * W) \left(\frac{X_i - X_j}{h} \right).$$

The use of the function Ω (see Definition 1.2) yields

$$\Omega(z) = (K * K * W * W - 2K * W * W + W * W)(z),$$

we obtain

$$\overline{\text{AISB}}\{\widehat{F}(\cdot, h)\} = -\frac{h}{n^2} \sum_{i,j=1}^n \Omega \left(\frac{X_i - X_j}{h} \right). \tag{3.13}$$

Integrating by parts and using Lemma 1.2 we evaluate useful moments of Ω, see Complements for details.

As in Chap. 2, $\overline{\text{AISB}}\{\widehat{F}(\cdot, h)\}$ is biased estimate of AISB$\{\widehat{F}(\cdot, h)\}$. The unbiased estimate takes the form

$$\widehat{\text{AISB}}\{\widehat{F}(\cdot, h)\} = -\frac{h}{n^2} \sum_{\substack{i,j=1 \\ i \neq j}}^n \Omega \left(\frac{X_i - X_j}{h} \right) \tag{3.14}$$

and

$$\widehat{\text{AIV}}^* \{\widehat{F}(\cdot, h)\} = \text{AIV}^* \{\widehat{F}(\cdot, h)\}.$$

In this context

$$\widehat{\text{AMISE}}^* \{\widehat{F}(\cdot, h)\} = \text{AIV}^* \{\widehat{F}(\cdot, h)\} + \widehat{\text{AISB}} \{\widehat{F}(\cdot, h)\}.$$

Taking the relations (3.7) and (3.14) into account we arrive at the formula for $\hat{h}_{\text{IT}}^F$

$$\hat{h}_{\text{IT}}^F = -\frac{4 \hat{h}_{\text{IT}}^F}{c_1 n} \sum_{\substack{i,j=1 \\ i \neq j}}^{n} \Omega \left(\frac{X_i - X_j}{\hat{h}_{\text{IT}}^F} \right). \tag{3.15}$$

and

$$\hat{h}_{\text{IT}}^F = \arg \min_{h \in H_n} \widehat{\text{AMISE}}^* \{\widehat{F}(\cdot, h)\}.$$

The solution of (3.15) is a fixed point of the function (denoted by φ) standing on the right hand side. Thus we consider one step iterative method:

$$\hat{h}_{\text{IT}, j+1}^F = \varphi(\hat{h}_{\text{IT}, j}^F), \ j = 0, 1, \ldots. \tag{3.16}$$

Since it would be difficult to verify whether the conditions for the convergence of the iterative process are satisfied we recommend to apply Steffensen's method (Stoer and Bulirsch (2002)). This method consists of the following steps:

$$
\begin{aligned}
\hat{h}_{\text{IT}, 0}^F &= \hat{h}_{\text{MS}}^F \\
t_j &= \varphi(\hat{h}_{\text{IT}, j}^F) \\
z_j &= \varphi(t_j) \\
\hat{h}_{\text{IT}, j+1}^F &= \hat{h}_{\text{IT}, j}^F - \frac{(t_j - \hat{h}_{\text{IT}, j}^F)^2}{z_j - 2t_j + \hat{h}_{\text{IT}, j}^F}.
\end{aligned}
\tag{3.17}
$$

According to our assumptions the optimal bandwidth $\hat{h}_{\text{IT}}^F$ is a simple fixed point, i.e., $\varphi'(\hat{h}_{\text{IT}}^F) \neq 1$. Then the method (3.17) provides a convergent sequence to the point $\hat{h}_{\text{IT}}^F$ (see Stoer and Bulirsch (2002) for details). The value $\hat{h}_{\text{MS}}^F$ given in (3.8) can be taken as a suitable initial approximation. This method is locally quadratically convergent method. Concerning the criterion for suitable approximation the relative error

$$\left| \frac{\hat{h}_{\text{IT}, j+1}^F - \hat{h}_{\text{IT}, j}^F}{\hat{h}_{\text{IT}, j}^F} \right| \leq \varepsilon,$$

$\varepsilon > 0$ is a given tolerance, can be recommended.

Statistical properties of estimates are described in the following theorem.

Theorem 3.2. *Let $F \in C^2$, F'' be square integrable, the kernel K belong to the class $S_{0,2}$ and $\lim\limits_{n \to \infty} h = 0$, $\lim\limits_{n \to \infty} nh = \infty$. Then*

$$
\begin{aligned}
E(\widehat{\mathrm{AISB}}\{\widehat{F}(\cdot, h)\}) &= \mathrm{AISB}\{\widehat{F}(\cdot, h)\} + o(h^4) \\
\mathrm{var}(\widehat{\mathrm{AISB}}\{\widehat{F}(\cdot, h)\}) &= \tfrac{2h^3}{n^2} V(\Omega) V(f) + o(n^{-2}h^3).
\end{aligned}
\tag{3.18}
$$

Proof. The proof is shifted to Complements. □

Corollary 3.2. *The relative convergence rate of $\hat{h}_{\mathrm{IT}}^F$ to h_{opt}^F can be expressed as*

$$
\frac{\hat{h}_{\mathrm{IT}}^F - h_{opt}^F}{h_{opt}^F} = O\left(n^{-\frac{1}{6}}\right).
$$

See Complements for the proof.

3.4 Boundary effects

The problem of boundary effects in kernel estimation was described more precisely in Sec. 2.7 for the case of kernel density estimation. For kernel distribution estimations the situation is similar. Although there is a vast literature on the boundary correction in the density estimation context, the boundary effects problem in the distribution function context has been less studied.

As in Chap. 2, we suppose the support of f (and the related distribution F) is $[0, \infty)$. We have denoted the interval $[0, h]$ as the "boundary region" and points from (h, ∞) as "interior" points. For more detailed description see Chap. 2.

The basic properties of $\widehat{F}(x, h)$ at interior points are well-known (*e.g.*, Lejeune and Sarda (1992), see also Sec. 3.2, formula (3.2)).

The performance of $\widehat{F}(x, h)$ at boundary points, *i.e.*, for $0 \le x \le h$, however, differs from the interior points due to boundary effects. More specifically, the bias$\{\widehat{F}(x, h)\}$ is of order $O(h)$ instead of $O(h^2)$ at boundary points, while the variance of $\widehat{F}(x, h)$ is of the same order. For more detailed discussion see, *e.g.*, Koláček (2008).

3.4.1 *Generalized reflection method*

In this paragraph, a kernel type estimator of the distribution function that removes boundary effects near the end points of the support is mentioned.

The estimator is based on a boundary corrected kernel estimator of distribution functions and it is based on ideas of Karunamuni and Alberts (2005a,b), and Zhang *et al.* (1999) developed for boundary correction in kernel density estimation. The basic technique of construction of the proposed estimator is a kind of a generalized reflection method involving reflecting a transformation of the observed data.

The proposed estimator takes the form

$$\widehat{F}_G(x,h) = \frac{1}{n} \sum_{i=1}^{n} \left\{ W\left(\frac{x - g_1(X_i)}{h} \right) - W\left(-\frac{x + g_2(X_i)}{h} \right) \right\}. \quad (3.19)$$

Functions g_1 and g_2 are two transformations that need to be determined. Some discussion on the choice of g_1 and g_2 and other various improvements can be found in Koláček and Karunamuni (2009). The trivial choice $g_1(y) = g_2(y) = y$ represents the "classical" reflection method estimator. In the mentioned paper, expressions for the bias and variance of the proposed estimator are derived. Some applications can be found in Horová *et al.* (2008a); Koláček and Karunamuni (2009, 2011).

3.5 Application to data

3.6 Simulations

The procedures described in the previous sections are applied to two data sets. The first one contains simulated data from Example 2.1. In this case the exact optimal bandwidth can be evaluated: $h_{opt,0,2} = 0.43121$ for the Epanechnikov kernel. The iterative method provides the best approximation of the optimal bandwidth: $\hat{h}_{IT}^F = 0.42153$, other methods give $\hat{h}_{PI}^F = 0.3365$ and $\hat{h}_{MS}^F = 0.6851$. The estimate of the distribution function (solid line) for the Epanechnikov kernel and $\hat{h}_{IT}^F$ together with the empirical distribution function (dashed line) are presented in Figure 3.5.

3.7 Application to real data

3.7.1 *Trout PCB data*

We estimate the distribution function for the bioaccumulation of PCB (Polychlorinated Biphenyl) in the tissue in Lake Ontario's trout (see also Horová *et al.* (2008a)). The data set consists of the concentrations of total

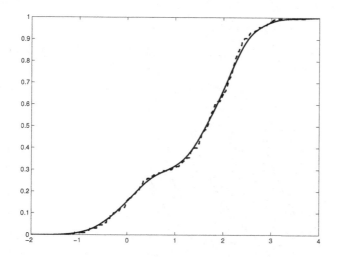

Fig. 3.5   Estimate of the distribution function for simulated data.

PCB in 177 fish caught in the Lake in the summer of 2001. These fish were classified according to their age and for each age the empirical and the smooth estimate of the distribution functions are shown in Figure 3.6. It is clear that:

- The smooth estimates of F provide very good fit for data.
- The distribution shifts to the right as the age increases indicating the increase of PCB level in older fish.

The information gained from this analysis can be important for environmental regulators where the interest is to place restrictions on fish consumption based on fish age.

3.8   Use of MATLAB toolbox

The toolbox can be downloaded from the web page
http://www.math.muni.cz/english/science-and-research/
developed-software/232-matlab-toolbox.html.

3.8.1   *Running the program*

The toolbox is launched by command `kscdf`. Launching without parameters will cause the start to the situation when only data input (button

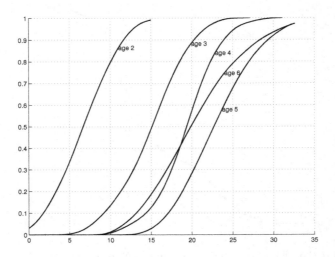

Fig. 3.6   Estimate of distribution functions for trout PCB data.

(1)) or terminating the program (button (2)) is possible (see Fig. 3.7). In the subroutine for data input (Fig. 3.8) you can choose the source

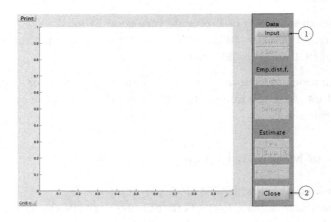

Fig. 3.7   Start of the program.

of the data – you can choose reading data from the workspace (buttons (3)), from the file (button (4)) or create the simulated data (button (5)). Then select the name of the variable in the list (6). If the true distribution

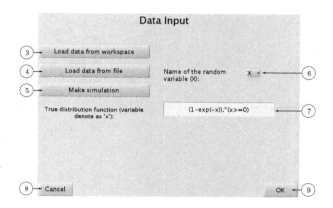

Fig. 3.8   Data input.

function is known (for simulated data, for instance), put it to the text field ⑦. You can use it to compare with the final estimate of the distribution function.

The data input can be cancelled by button ⑧ or confirmed by button ⑨.

3.8.2   *Main figure*

After data input you can view data (button ⑩, Fig. 3.9) and save chosen values (button ⑪).

Button ⑫ displays the empirical distribution function (Fig. 3.10). Button ⑬ invokes the setting of the parameters for smoothing (Fig. 3.11).

3.8.3   *Setting the parameters*

In this subroutine you can choose a predefined kernel (button ⑯), an optimal kernel (button ⑰) and draw the kernel (button ⑱), but only kernels of order $\nu = 0, k = 2$ are available. The bandwidth can be selected in box ⑲ if the kernel is selected. Four methods for bandwidth choice are implemented (see Sec. 3.3). The points for drawing the estimate of the density can be set up (㉑) in the right part of the window. Finally, you can confirm

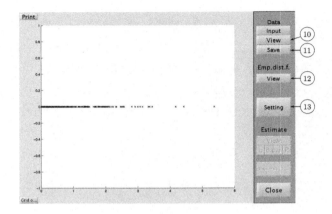

Fig. 3.9   Data view.

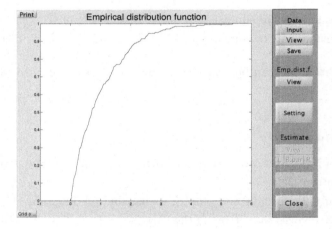

Fig. 3.10   Empirical distribution function.

the setting (button ㉒) or close the windows without change of the parameters (button ㉓).

3.8.4   Eye-control method

You can use the *"Eye-control"* method (see Fig. 3.12) for bandwidth choice
by button ⑳. First set the initial bandwidth and the step for its increasing or decreasing in boxes ㉔ and ㉕. Pressing the middle button from

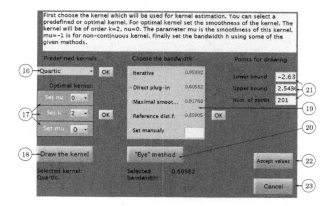

Fig. 3.11 Setting the parameters.

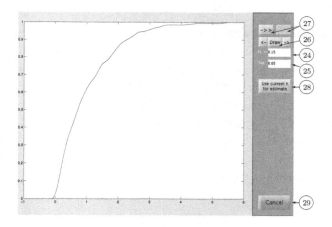

Fig. 3.12 "Eye-control" method.

㉖ displays the estimate for actual bandwidth, the arrows at the left and the right side cause increasing or decreasing the bandwidth by a step and redrawing the figure. By buttons ㉗ you can run and stop the gradual increasing the bandwidth and drawing the corresponding estimate. Finally it is possible to accept selected bandwidth (button ㉘) or cancel the procedure (button ㉙) .

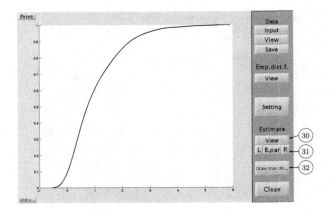

Fig. 3.13   Kernel estimate.

3.8.5   *The final estimation*

If the kernel and the bandwidth are set the estimate is displayed (button
(30), Fig. 3.13). Now, it is possible to set the correction of the boundary
effects (the middle button in (31)) and display the result by pressing "L" or
"R" for the left or the right boundary (see Fig. 3.14). We can also display
the true distribution function (button (32), Fig. 3.14), if it was entered.

The grid and the separate window with the estimate is also at disposal
(buttons (33) and (34)).

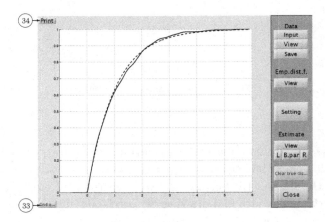

Fig. 3.14   Estimate with boundary correction and the true distribution function.

3.9 Complements

Proof of Theorem 3.1 (see also Lejeune and Sarda (1992); Nováková (2009)).

Proof. We start with the evaluation of $E\widehat{F}(x,h)$ at the point $x \in \mathbb{R}$:

$$E\widehat{F}(x,h) = \int\limits_{-\infty}^{\infty} W\left(\frac{x-y}{h}\right) f(y)dy$$

$$= h \int\limits_{-\infty}^{1} W(t)f(x-ht)dt + h \int\limits_{1}^{\infty} W(t)f(x-ht)dt.$$

The first integral is denoted as I_1, the second one as I_2.

Let us calculate I_1. The use of properties of the function W and integration by parts lead to

$$I_1 = -F(x-h) + \int\limits_{-1}^{1} K(t)F(x-ht)dt. \tag{3.20}$$

Further, we use the Taylor expansion of the form

$$F(x-ht) = F(x) - htF'(x) + \frac{h^2t^2}{2}F''(x) + o(h^2)$$

and substitute it in (3.20). Hence

$$I_1 = -F(x-h) + F(x) + \frac{1}{2}F''(x)h^2\beta_2 + o(h^2).$$

As concerns I_2 it is easy to see that

$$I_2 = F(x-h). \tag{3.21}$$

Here, we notice that (3.20) and (3.21) yield an interesting formula for bias

$$E\widehat{F}(x,h) - F(x) = \int\limits_{-1}^{1} K(t)F(x-ht)dt - F(x).$$

Putting I_1 and I_2 together we obtain

$$\text{bias}\{\widehat{F}(x,h)\} = \frac{1}{2}F''(x)h^2\beta_2 + o(h^2). \tag{3.22}$$

The proof of the variance term is slightly more complicated than the proof of the bias term. According to the definition

$$\text{var}\{\widehat{F}(x,h)\} = \frac{1}{n}\left[EW^2\left(\frac{x-X}{h}\right) - E^2W\left(\frac{x-X}{h}\right)\right].$$

We are only dealing with the first term, since the second is given in (3.22). Thus

$$I_3 = \int_{-\infty}^{\infty} W^2\left(\frac{x-y}{h}\right) f(y)dy$$

$$= h\int_{-\infty}^{1} W^2(t)f(x-ht)dt + h\int_{1}^{\infty} f(x-ht)dt$$

$$= h\int_{-1}^{1} W^2(t)f(x-ht)dt + F(x-h).$$

Further, the integration by parts yields

$$I_3 = 2\int_{-1}^{1} W(t)W'(t)F(x-ht)dt$$

$$= 2\int_{-1}^{1} W(t)W'(t)\{F(x) - htF'(x) + o(h)\}dt$$

$$= 2F(x)\int_{-1}^{1} W(t)W'(t)dt - 2hf(x)\int_{-1}^{1} tW(t)W'(t)dt + o(h).$$

The use of facts 3° and 4° leads to

$$I_3 = F(x) - hf(x)\left(1 - \int_{-1}^{1} W^2(t)dt\right) + o(h). \qquad (3.23)$$

Since (3.22) can be expressed as

$$EW\left(\frac{x-X}{h}\right) = F(x) + o(h), \qquad (3.24)$$

then (3.23) and (3.24) provide

$$\text{var}\{\widehat{F}(x,h)\} = \frac{1}{n}F(x)(1-F(x)) - \frac{1}{n}hf(x)\left(1 - \int_{-1}^{1} W^2(t)dt\right) + o\left(\frac{h}{n}\right).$$

Thus

$$\text{MSE}\{\widehat{F}(x,h)\} = \frac{1}{n}F(x)(1-F(x)) - \frac{1}{n}hf(x)\left(1 - \int_{-1}^{1} W^2(t)dt\right)$$

$$+ \frac{1}{4}(F''(x))^2 h^4\beta_2^2 + o\left(\frac{h}{n} + h^4\right)$$

and

$$\text{MISE}\{\widehat{F}(\cdot, h)\} = \int \text{MSE}\{\widehat{F}(x, h)\} dx$$

and thus the formula (3.3) follows. ☐

Using Lemma 1.2 we can evaluate useful moments of Ω.

Lemma 3.2. *Let* $K \in S_{0,2}$, *then*

$$\int z^j \Omega(z) dz = 0, \quad j = 0, 1$$

$$\int z^2 \Omega(z) dz = \frac{\beta_2^2}{2}.$$

Proof of Theorem 3.2:

Proof. We give only a sketch of this proof, since it is very similar to that given in density estimates. We need to evaluate $E(\widehat{\text{AISB}}\{\widehat{F}(\cdot, h)\})$:

$$E(\widehat{\text{AISB}}\{\widehat{F}(\cdot, h)\}) = -\frac{h}{n^2} E \sum_{\substack{i,j=1 \\ i \neq j}}^{n} \Omega\left(\frac{X_i - X_j}{h}\right)$$

$$= -h\left(1 - \frac{1}{n}\right) E\, \Omega\left(\frac{X_1 - X_2}{h}\right)$$

$$= -h^2\left(1 - \frac{1}{n}\right) \int \int \Omega(u) f(y - uh) f(y) du\, dy$$

$$= -h^2\left(1 - \frac{1}{n}\right) \int f(y) \int \Omega(u) \times$$

$$\times \left\{ f(y) - uhf'(y) + \frac{u^2 h^2}{2} f''(y) + o(h^2) \right\} du\, dy.$$

Using Lemma 3.2 leads to

$$E(\widehat{\text{AISB}}\{\widehat{F}(\cdot, h)\}) = -h^2\left(1 - \frac{1}{n}\right) \left\{ \frac{\beta_2^2 h^2}{4} \int f(y) f''(y) dy + o(h^2) \right\}.$$

By integrating $\int f(y) f''(y) dy$ and neglecting asymptotic small terms we obtain

$$E(\widehat{\text{AISB}}\{\widehat{F}(\cdot, h)\}) = \frac{\beta_2^2 h^4}{4} \int (F''(y))^2 dy + o(h^4)$$

$$= \text{AISB}\{\widehat{F}(\cdot, h)\} + o(h^4).$$

Further, proceed in a similar way as in the case of density estimates we arrive at the variance expression. ☐

Corollary 3.2 can be proved in the same way as in the case of univariate density derivative estimation. For a detailed proof see Complements in Chap. 2.

Chapter 4

Kernel estimation and reliability assessment

Consider the following situation. Each of a set of objects is known to belong to one of two classes. An assignment procedure assigns each object to a class on the basis of information observed about this object. But such a procedure need not to be perfect and errors could be made and the object is assigned to an incorrect class. Thus it is necessary to evaluate the quality of performance of the procedure. There are many possible ways to measure the performance of the classification rules. It is often very helpful to have a way of displaying and summarizing performance over a wide range of conditions. This aim is fulfilled by ROC (Receiver Operating Characteristic) curve. It is a single curve summarizing the distribution functions of the scores of two classes.

The name "Receiver Operating Characteristic" curve comes from the notion that given the curve we – receivers of the information – can use or operate at any point on the curve by using the appropriate decision threshold (cutoff point) – characteristic.

They were originally developed for electronic – signal detection theory (see Zhou *et al.* (2002) for details). In 1951 the concept of ROC curves was introduced to medicine by Lusted (1971). Recently, there has been an increased use of ROC curves for accessing the effectiveness of a continuous diagnostic test in distinguishing between diseased and healthy individuals. There is a large number of research papers and monographs. We recommend excellent monographs by Zhou *et al.* (2002); Pepe (2004); Krzanowski and Hand (2009) for very detailed reading and for a bibliography therein. With respect to this fact we will give only a brief survey on ROC curves and focus on some indices connected with them, on application in medicine, economy, finance and mainly on implementation of all discussed methods in MATLAB.

4.1 Basic Definition

Let $\mathcal{G}_1$ be a group of objects with condition (*e.g.*, diseased individuals) and $\mathcal{G}_0$ be a group of objects without condition (*e.g.*, healthy individuals). Let one-dimensional absolute continuous random variable X be a diagnostic test variable. We use this variable to test whether the object falls into group $\mathcal{G}_0$ or $\mathcal{G}_1$, *i.e.*, for a given cutoff point $c \in \mathbb{R}$ the object is classified as $\mathcal{G}_1$ if $X \geq c$ and as $\mathcal{G}_0$ otherwise. By convention, large values of X are considered more indicative of disease.

We aim to evaluate whether a given test is able to correctly detect the presence or the absence of the condition. We illustrate these notions by an example.

Example 4.1. Let groups $\mathcal{G}_0$ and $\mathcal{G}_1$ be given – see Table 4.1.

Table 4.1 Test variable for groups $\mathcal{G}_0$ and $\mathcal{G}_1$.

$\mathcal{G}_0$:							
2.0563	0.5835	0.9751	-0.7482	-0.3314	-0.1927	0.7767	-0.1965
0.4961	-1.4509	-0.2502	-0.5824	-1.1151	0.3592	0.1994	0.0237
-0.9431	0.7973	0.2476	-0.2115	0.0162	-0.1853	-1.2376	-0.2020
-0.5879	-0.6924	-0.8177	-0.3773	-1.4161	0.6818	0.3677	-0.0142
-0.0246	-0.5644	0.7203	-0.0942	-0.5052	0.9556	-0.1589	-0.4165
-0.2077	-0.5996	-0.7921	1.7862	1.1706	0.2175	-0.8889	-0.6120
-0.1248	0.5596	-0.9419	-1.6475	-1.0247	0.2358	0.2767	0.3194
-0.0921	0.1299	-0.3367	0.6095				
$\mathcal{G}_1$:							
0.9358	3.6035	3.2347	1.7704	0.4938	1.5554	1.8441	2.2761
1.7388	2.4434	2.3919	0.7493	1.0520	1.2589	1.4922	1.6794
2.0125	-1.0292	1.5430	3.2424	0.9333	2.9337	2.3503	1.9710
2.1825	0.4349	1.9155	3.6039	2.0983	2.0414	1.2658	1.9692
2.2323	2.4264	1.6272	1.7635	4.0237	-0.2584	4.2294	2.3376

Figure 4.1 presents the data points of both groups. Consider a cutoff point $c = 1$. Then any element A of Table 4.1 is classified as $\mathcal{G}_1$ if $A \geq c$. Otherwise, it is classified as $\mathcal{G}_0$. It can bee seen that some objects are classified incorrectly.

First, we introduce some notation. We use the binary variable D to denote the presence of the condition:

$$D = \begin{cases} 1 & \text{for presence of condition} \\ 0 & \text{for absence of condition} \end{cases}$$

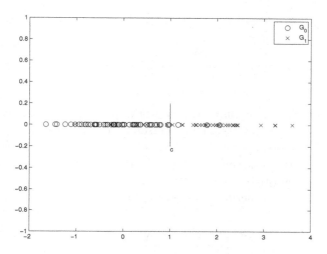

Fig. 4.1   Data groups $\mathcal{G}_0$ (○ symbols) and $\mathcal{G}_1$ (× symbols).

The variable $\mathcal{T}$ is the result of the diagnostic test:

$$\mathcal{T} = \begin{cases} 1 & \text{positive test result} \\ 0 & \text{negative test result} \end{cases}$$

The results can be summarized in a *confusion matrix* (see Table 4.2, also known as a *decision matrix, contingency table* or *count table*) where characteristics are defined as follows:

True positive: the test correctly indicates the presence of the condition, TP denotes a number of these objects.

True negative: the test correctly indicates the absence of the condition, TN denotes a number of these objects.

False negative: the test falsely indicates the absence of the condition though the condition is truly present, FN denotes a number of these objects.

False positive: the test falsely indicates the presence of the condition though it is not truly present, FP denotes a number of these objects.

The test has two types of errors, *false positive* error and *false negative* error. An ideal test has no false positives and no false negatives. False negative errors, *i.e.*, missing disease, that is present, can result in people foregoing needed treatment for their disease. The consequence can be serious as death. False positive errors tend to be less serious. People without

Table 4.2   Test results (confusion matrix)

	$\mathcal{T} = 1$	$\mathcal{T} = 0$
$(D = 1)$	True positive	False negative
$(D = 0)$	False positive	True negative

disease are subjected to unnecessary work-up procedures or even treatment. The negative impacts include personal inconvenience and stress but the long-term consequences are usually relative minor. Clearly, for a good performance of the test we require high "true" and low "false".

The test diagnostic *accuracy* is the ability of the test to detect correctly a condition when it is actually present and to correctly rule out when it is truly absent:

$$accuracy = \frac{TN + TP}{N},$$

where $N = TP + FN + FP + TN$ is the total number of objects.

In medical research the *sensitivity* and *specificity* of the test are often used to describe test performance.

The *sensitivity* (Se) of the test is its ability to detect the condition when it is present.

$Se = P(\mathcal{T} = 1|D = 1) = TP/(TP+FN)$ is a probability P that the test result is positive $(\mathcal{T} = 1)$, given that the condition is present $(D = 1)$.

The *specificity* (Sp) of the test is its ability to exclude the condition when it is absent.

$Sp = P(\mathcal{T} = 0|D = 0) = TN/(FP+TN)$ is a probability P that the test result is negative $(\mathcal{T} = 0)$, given that the condition is absent $(D = 0)$.

The accuracy of the test is often summarized in a *Receiver Operating Characteristic (ROC) curve*. The ROC curve is a method of describing the intrinsic accuracy of a test apart from the given cutoff point c.

Definition 4.1. The ROC curve is defined as a plot of the probability of false classification $(1 - Sp)$ of subjects from $\mathcal{G}_0$ versus the probability of true classification (Se) of subjects from $\mathcal{G}_1$ across all possible cutoff point values of c.

Each point on the graph is generated by a different cutoff point c. Figure 4.2 presents the points of the ROC curve for data from Example 4.1. We use line segments to connect the points generated by given cutoff points $c = -2, -1, 0, 1, 2, 3, , 4$, successively.

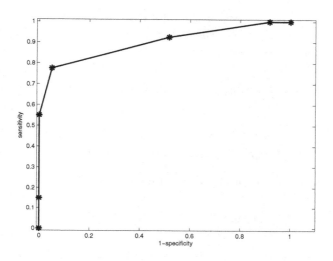

Fig. 4.2   ROC curve.

Now, we derive an explicit formula for the ROC curve. Let $\mathbf{Y}$ be an m-dimensional random vector. Further, let one-dimensional absolute continuous random variable $X = X(\mathbf{Y}) \in \mathbb{R}$ be a diagnostic test variable for the outcome vector $\mathbf{Y}$ with the density function

$$f(x) = p_0 f_0(x) + p_1 f_1(x), \quad p_0 > 0, \ p_1 > 0, \ p_0 + p_1 = 1$$

and F_0 and F_1 are distribution functions of groups $\mathcal{G}_0, \mathcal{G}_1$, respectively. Evidently, f_0, f_1 are conditioned densities of X, given $D = 0, 1$.

For a given cutoff point $c \in \mathbb{R}$

$$F_0(c) = P(X \le c|\mathcal{G}_0) = \int_{-\infty}^{c} f_0(x)dx = Sp$$

and

$$F_1(c) = P(X \le c|\mathcal{G}_1) = \int_{-\infty}^{c} f_1(x)dx \Rightarrow Se = 1 - F_1.$$

Thus F_0 is the specificity (Sp) of the test and $1 - F_1$ is the sensitivity (Se) of the test and the ROC curve is displayed by plotting $1 - F_1(c)$ against $1 - F_0(c)$ for a range of cutoff points $c \in \mathbb{R}$.

Figure 4.3 shows the above mentioned characteristics for a given cutoff point c and densities f_0 and f_1.

Let p be the probability of the false classification of subjects from $\mathcal{G}_0$ and q the probability of the true classification of subjects from $\mathcal{G}_1$.

As the first, we evaluate p:

$$p = \int_c^\infty f_0(x)dx = 1 - F_0(c) \Rightarrow c = F_0^{-1}(1-p),$$

where

$$F_0^{-1}(p) = Q_0(p) = \inf\{\theta \in \mathbb{R} : F_0(\theta) \geq p\}$$

is the quantile function of F_0. Then, using this result we evaluate q:

$$q = \int_c^\infty f_1(x)dx = 1 - F_1(c) = 1 - F_1\left(F_0^{-1}(1-p)\right).$$

Thus, we arrive at the explicit formula for the ROC curve

$$\mathrm{ROC}(p) = 1 - F_1\left(F_0^{-1}(1-p)\right). \tag{4.1}$$

Misclassification rate of the test will be seen from the distance between the ROC curve and the upper left corner $[0,1]$ (see Fig. 4.6). Special cases of the ROC curves are:

A *perfectly accurate* – all objects are classified correctly (see Fig. 4.4).

A *perfectly inaccurate* – objects with the condition are classified incorrectly as negative and patient without condition ale classified incorrectly as positive.

A *diagonal, chance diagonal* or *a curve for random model* – the test is not usable for separation of objects (see Fig. 4.5).

Remark 4.1. The ROC curve is usually concave (see Zhou *et al.* (2002) or Lloyd (2002) for convex ROC curves).

Example 4.2. Consider densities f_0 and f_1 for groups $\mathcal{G}_0$ and $\mathcal{G}_1$

$$f_0(x) = \frac{1}{\sqrt{\pi}} e^{-(x+1)^2}, \qquad f_1(x) = \frac{1}{0.5\sqrt{2\pi}} e^{-\frac{(x-1)^2}{2 \cdot 0.5^2}}.$$

In this case the ROC curve is close to the perfectly accurate one, see Fig. 4.6.

Notation 4.1. Let $X_j(j = 0,1)$ denote a random variable X if $D = j$, $j = 0,1$ (X_0, X_1 are independent).

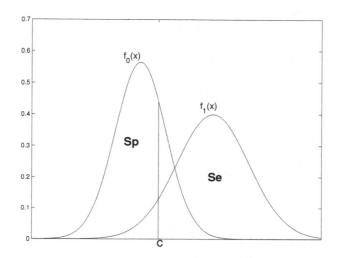

Fig. 4.3   Characteristics for the given cutoff point c.

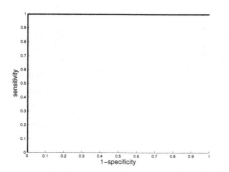

Fig. 4.4   Perfectly accurate test: sensitivity is 1 when 1-specificity is 0.

Fig. 4.5   Diagonal – chance diagonal (random model).

Remark 4.2. The function $\text{ROC}(p) = R(p)$ also possesses an important property – it is the distribution function of $1 - F_0(X_1)$. Put $V = 1 - F_0(X_1)$, then

$$
\begin{aligned}
F_V(p) = P(V \leq p) &= P(1 - F_0(X_1) \leq p) \\
&= P(X_1 \geq F_0^{-1}(1 - p)) \\
&= 1 - F_1(F_0^{-1}(1 - p)) = R(p).
\end{aligned} \tag{4.2}
$$

Thus $R(p)$ is the nonzero distribution function of the p-value $1 - F_0(X_1)$ for testing the null hypothesis that an individual comes from $\mathcal{G}_0$.

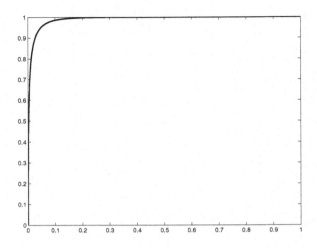

Fig. 4.6   ROC curve for the model from Example 4.2.

4.2   Estimation of ROC curves

In practice the population distributions are never fully specified. The parametric model could be a poor approximation of the true population model. The nonparametric estimators of ROC curves seem appealing in many situations where limited or no information is available about the population distribution of the diagnostic variables. Thus, we only briefly mention the parametric model (see DeLong *et al.* (1988); Pepe (2004); Zhou *et al.* (2002) for details and analysis).

4.2.1   *Binormal model*

A simple parametric approach is to assume that the density function of diagnostic test variable X is the mixture of two normal (gaussian) components with means μ_0, μ_1 and variances σ_0^2 and σ_1^2, respectively:

$$f_j(x) = \frac{1}{\sigma_j\sqrt{2\pi}} e^{-\frac{(x-\mu_j)^2}{2\sigma_j^2}}, \quad j = 0, 1\,.$$

Then the ROC curve has the form

$$R(p) = \Phi\left(a + b\Phi^{-1}(p)\right),$$

where

$$a = \frac{\mu_1 - \mu_0}{\sigma_1}, \quad b = \frac{\sigma_0}{\sigma_1}, \quad 0 \le p \le 1$$

and Φ is the standard normal distribution function, $\Phi(x) = \frac{1}{\sqrt{2\pi}} \int\limits_{-\infty}^{x} e^{-\frac{t^2}{2}} dt.$

The parameters a, b are called *separation* and *symmetry* parameters, respectively.

An elliptically contoured distribution (ECD) is a natural generalization of the multivariate normal case (see Fang *et al.* (1990) for details). In the paper Forbelská (2007) the attention is paid to the case when the density function f is a mixture of two ECD components.

4.2.2 Nonparametric estimates

Due to the reasons mentioned above the attention will be turned to non-parametric methods, which are applicable very generally.

The simplest nonparametric estimation method for the ROC curve, the empirical ROC curve, involves replacing F_0 and F_1 by their empirical distribution functions

$$\widetilde{F}_j(x) = \sum_{i=1}^{n_j} I_{(-\infty, x]}(X_{ji}), \quad j = 0, 1,$$

where $X_{01}, \ldots, X_{0,n_0}$ and $X_{11}, \ldots, X_{1,n_1}$ are independent samples from F_0 and F_1, respectively.

Hence, the empirical estimation

$$\widehat{R}_{emp}(p) = 1 - \widetilde{F}_1(\widetilde{F}_0^{-1}(1-p)), \quad 0 \le p \le 1$$

is the nonparametric estimator of $R(p)$. This estimator is a step function on the unit square. One major weakness of the empirical ROC curve is its jagged form. Zou *et al.* (1997) described a kernel method to estimate a smooth ROC curve from continuous data (see also Zhou *et al.* (2002)). They suggested estimating the points on the ROC curve

$$\left(\bar{F}_0(c), \bar{F}_1(c) \right)$$

through the integral of the densities, and the densities are estimated by

$$\hat{f}_j(x, h_j) = \frac{1}{n_j h_j} \sum_{i=1}^{n_j} K\left(\frac{x - X_{ji}}{h_j} \right), \quad j = 0, 1,$$

where $K(x) = \frac{15}{16}(1 - x^2)^2 I_{[-1,1]}$ and the bandwidth

$$h_j = 0.9 \min(\hat{\sigma}_{SD}, \sigma_{IQR}/1.34) n_j^{-1/5}, \quad j = 0, 1$$

(for $\hat{\sigma}_{SD}$ and $\hat{\sigma}_{IQR}$ see (2.13) and (2.14)).
Thus

$$\bar{F}_j(c) = \frac{1}{n_j h_j} \sum_{i=1}^{n_j} \int\limits_{-\infty}^{c} K\left(\frac{x - X_{ji}}{h_j}\right) dx, \quad j = 0, 1.$$

These integrals can be evaluated by means of the trapezoidal rule. However, the choice of bandwidths is not optimal because the optimal bandwidths for densities ($h^f = O(n^{-1/5})$) do not imply the optimality for distribution functions ($h^F = O(n^{-1/3})$).

Lloyd (1998) therefore considered the direct estimation of F_0 and F_1 using kernel methods (see also Lloyd and Yong (1999)). In the paper Zhou and Harezlak (2002) four kernel smoothing methods for estimating the ROC curve have been compared. The result from the simulation study suggested that the kernel smoothing originally proposed by Altman and Léger (1995) was the best choice for estimation of the ROC curve (see Chap. 3).

Nevertheless, the resulting estimate still has some drawbacks (see *e.g.* Krzanowski and Hand (2009)). The estimate of F_0 and F_1 may be asymptotically optimal, there is no guarantee of asymptotic optimality for the ROC curve itself. Further, since F_0 and F_1 are estimated separately, the final ROC curve estimator is not invariant under a monotone transformation of the data. The papers by [Pepe (2004)] and Ren *et al.* (2004) approached this drawback by using local linear smoothing and penalized regression splines applied to linear model, respectively. Wan and Zhang (2008) proposed a semiparametric kernel distribution function estimator, based on which a new smooth semiparametric estimator for the ROC curve is constructed. Despite these shortcomings we will pay attention to kernel estimates because we concentrate mainly on indices and applications in medicine, economy, finance *etc.* and in such cases the kernel distribution function estimators on which the estimate $\widehat{R}(p)$ is based perform reasonably well. Moreover, we propose a method to overcome the boundary effects (see Koláček and Karunamuni (2011) and Sec. 3.4).

Kernel estimate of ROC curve

Let $\widehat{F}_0$ and $\widehat{F}_1$ be the kernel estimates of distribution functions of samples $X_{01}, \ldots, X_{0n_0}$ and $X_{11}, \ldots, X_{1n_1}$, respectively:

$$\widehat{F}_j(x, h_j) = \frac{1}{n_j} \sum_{i=1}^{n_j} W\left(\frac{x - X_{ji}}{h_j}\right), \quad j = 0.1 \qquad (4.3)$$

(see Chap. 3).

The corresponding kernel estimator of $R(p)$ is

$$\widehat{R}(p) = 1 - \widehat{F}_1\left(\widehat{F}_0^{-1}(1-p)\right), \quad 0 \le p \le 1. \tag{4.4}$$

The smooth estimator requires the choice of kernel K and two bandwidths. As concerns the kernel, the biweight kernel $K(x) = \frac{15}{16}(1-x^2)^2 I_{[-1,1]}$ is recommended.

Further, $R(p)$ can be considered as a distribution function of $1 - F_0(X_1)$ (see Remark 4.1) and thus $\widehat{R}(p)$ can be expressed by means of kernel estimate of a distribution function

$$\widehat{R}_I(p) = \frac{1}{n_1} \sum_{i=1}^{n_1} W\left(\frac{p - (1 - \widehat{F}_0(X_{1i}, h_0))}{h_1}\right). \tag{4.5}$$

The choice of bandwidths h_0 and h_1 has been treated especially by Lloyd (1998); Lloyd and Yong (1999). They also proved the following form of AMSE:

$$\text{AMSE}\left\{\widehat{R}(p)\right\} = \text{AV}(\widehat{R}(p)) + \text{AB}^2(\widehat{R}(p)), \tag{4.6}$$

where

$$\text{AV}(\widehat{R}(p)) = \frac{R(p)(1-R(p))}{n_1} + \frac{p(1-p)R'^2(p)}{n_0} + O\left(\frac{h_1}{n_1}\right),$$

$$\text{AB}(\widehat{R}(p)) = \frac{R''(p)}{2}\left\{h_0^2 f_0^2(\theta_p) + \frac{p(1-p)}{n_0}\right\} + \frac{1}{2}(h_0^2 - h_1^2)f_0'(\theta_p)$$

with $\theta_p = f_0^{-1}(1-p)$.

The optimal bandwidths for $\widehat{R}(p)$ are of order $h_i = O(n_i^{-1/3})$, $i = 0, 1$. They are the same as the optimal bandwidths for estimating F_i however the constants are not the same. For simplicity, bandwidths optimal for estimation F_i are usually employed.

At the end of this section we notice an interesting relation between optimal bandwidths for F_0 and F_1. At first, let $\hat{h}_{MS}^{F_0}$ and $\hat{h}_{MS}^{F_1}$ be bandwidths obtained by maximal smoothing principle (see Sec. 3.3.2). Then

$$\frac{\hat{h}_{MS}^{F_0}}{\hat{h}_{MS}^{F_1}} = \frac{\hat{\sigma}_0}{\hat{\sigma}_1}\left(\frac{n_1}{n_0}\right)^{1/3}. \tag{4.7}$$

Therefore, it is reasonable to choose h_0 and h_1 such that

$$\frac{h_1}{h_0} \approx \left(\frac{n_0}{n_1}\right)^{1/3}.$$

Table 4.3   Summary indices for the ROC curve

Index name	Notation	Definition
Area under the curve	AUC	$\int_0^1 R(p)dp$
Gini index	Gini	$2\mathrm{AUC} - 1$
Specific ROC point	$R(t_0)$	$R(t_0)$
Partial area under the curve	PAUC(t_0)	$\int_0^{t_0} R(p)dp$
Symmetry point	Sym	$R(Sym) = Sym$
Maximal improvement of sensitivity (Kolmogorov–Smirnov)	MIS	$\sup \lvert R(p) - p \rvert$

4.3   Summary indices based on the ROC curve

Numerical indices for the ROC curve are often used to summarize the curves. When it is not feasible to plot the ROC curve itself, such summary measures convey information about the curve. Summary indices are particularly useful when many tests are under consideration. Moreover, they can be used as the basic inferential statistics for the comparison of ROC curves. Table 4.3 brings some useful summary indices (Pepe (2004)).

4.3.1   *Area under the ROC curve*

The most commonly used global index of diagnostic accuracy is the area under the ROC curve denoted AUC. The formal definition is

$$\mathrm{AUC} = \int_0^1 R(p)dp. \tag{4.8}$$

The AUC takes a value between 0 and 1. Values close to 1 indicate that the test has the high diagnostic accuracy.

The AUC can be interpreted as a probability that a pair of individuals known to be from different groups will be correctly classified (Zhou *et al.* (2002)). A simple calculation shows that the AUC is exactly equal to the probability that X_0 is less than X_1:

$$\mathrm{AUC} = P(X_0 < X_1) \tag{4.9}$$

(see Krzanowski and Hand (2009)).

AUC for the binormal model

It is easy to show that for binormal model (Sec. 4.2.1)

$$\text{AUC} = \Phi\left(\frac{a}{\sqrt{1+b^2}}\right), \qquad (4.10)$$

Φ – standard normal distribution function.

An overview of nonparametric methods of AUC estimates follows.

Empirical AUC

The empirical AUC ($\widetilde{\text{AUC}}_{emp}$) is given by evaluating the trapezoidal area under each vertical slice of an empirical ROC curve having a straight-line segment as its top and then sum all individual areas, *i.e.*,

$$\widetilde{\text{AUC}}_{emp} = \frac{1}{n_0 n_1} \sum_{i=1}^{n_1} \sum_{j=1}^{n_0} U(X_{0j}, X_{1i},),$$

$$U(X_{0j}, X_{1i}) = \begin{cases} 1 & X_{1i} > X_{0j}, \\ \frac{1}{2} & X_{1i} = X_{0j}, \\ 0 & \text{otherwise.} \end{cases} \quad \begin{array}{l} j = 1, \ldots, n_0 \\ i = 1, \ldots, n_1. \end{array}$$

Remark 4.3. This is an analogy to the Mann-Witney U-statistics.

Composite trapezoidal rule

The estimates of F_0 and F_1 are evaluated in a suitable set

$$\{x_r \in \mathbb{R}; \ r = 0 \ldots N\}, \ \text{mostly} \ x_r = x_0 + rt, \ t \in \mathbb{R}, t > 0.$$

The kernel estimate $\widehat{R}$ of the ROC curve is formed by points $[p_r, \widehat{R}(p_r)]$ where

$$p_r = 1 - \widehat{F}_0(x_r), \quad \widehat{R}(p_r) = 1 - \widehat{F}_1(x_r), \quad r = 0, \ldots, N.$$

The composite trapezoidal rule yields

$$\widehat{\text{AUC}} = \sum_{r=1}^{N} \frac{1}{2}(p_{r-1} - p_r)\left(\widehat{R}(p_{r-1}) + \widehat{R}(p_r)\right) =$$

$$= \frac{1}{2} \sum_{r=1}^{N} \left(\widehat{F}_{0,h_0}(x_r) - \widehat{F}_{0,h_0}(x_{r-1})\right)\left(2 - \widehat{F}_{1,h_1}(x_{r-1}) - \widehat{F}_{1,h_1}(x_r)\right)$$

The trapezoidal rule underestimates systematically the AUC because all the points $[p_r, \widehat{R}(p_r)]$, $r = 0, \ldots, N$ are connected with straight-lines instead of a smooth concave curve.

Kernel estimate of AUC – method 1

In terms of distribution function the AUC can be expressed as

$$\text{AUC} = P(X_0 < X_1) = P(X_0 - X_1 < 0) = F_{X_0 - X_1}(0) = F^Z(0),$$

where $F^Z = F_{X_0 - X_1}$ is a distribution function of a random variable $Z = X_0 - X_1$. Then a kernel estimate of F^Z is

$$\widehat{F}^Z_{h_0,h_1}(x) = \frac{1}{n_0 n_1} \sum_{i=1}^{n_1} \sum_{j=1}^{n_0} W\left(\frac{x - (X_{0j} - X_{1i})}{\sqrt{h_0^2 + h_1^2}}\right),$$

where h_0 and h_1 are the bandwidths for F_0 and F_1, respectively (Lloyd (1998)). Hence the kernel estimate $\widehat{\text{AUC}}_I$ of AUC is given by

$$\widehat{\text{AUC}}_I = \widehat{F}^Z_{h_0,h_1}(0) = \frac{1}{n_0 n_1} \sum_{i=1}^{n_1} \sum_{j=1}^{n_0} W\left(\frac{X_{1i} - X_{0j}}{\sqrt{h_0^2 + h_1^2}}\right).$$

Kernel estimate of AUC – method 2

We can estimate F^Z by means of the only bandwidth h, by the formula

$$\widehat{F}^Z_h(x) = \frac{1}{n_0 n_1} \sum_{i=1}^{n_1} \sum_{j=1}^{n_0} W\left(\frac{x - (X_{0j} - X_{1i})}{h}\right)$$

and

$$h^{F^Z}_{opt} = (n_0 n_1)^{-1/3} \left(\frac{c_1}{-\beta_2^2 \psi_1^Z}\right)^{1/3}, \qquad h^{F^Z}_{opt} \approx O((n_0 n_1)^{-1/3}),$$

where

$$\psi_1^Z = \int \left(F^{Z''}(x)\right)^2 dx,$$

(see Chap. 3). Then

$$\widehat{\text{AUC}}_{II} = \widehat{F}^Z_h(0) = \frac{1}{n_0 n_1} \sum_{i=1}^{n_1} \sum_{j=1}^{n_0} W\left(\frac{X_{1i} - X_{0j}}{h}\right), \quad h = \hat{h}^{F^Z}_{opt}.$$

In this case we have to choose only one smoothing parameter and therefore this approach possesses good asymptotic properties (see Chap. 3).

Kernel estimate of AUC – method 3

This method uses the $\widehat{R}_I(p)$ of formula (4.5) for the ROC curve estimate. Direct integration gives:

$$\widehat{\mathrm{AUC}}_{III} = \int\limits_0^1 \widehat{R}(p)\,dp = \int\limits_0^1 \widehat{F}_{V,h_1}(p)\,dp$$

$$= \frac{1}{n_1}\sum_{i=1}^{n_1}\int\limits_0^1 W\left(\frac{p-(1-\widehat{F}_0(X_{1,i}))}{\tilde{h}_1}\right)dp,$$

where

$$\widehat{F}_{0,h_0}(X_{1,i}) = \frac{1}{n_0}\sum_{j=1}^{n_0} W\left(\frac{X_{1,i}-X_{0,j}}{h_0}\right).$$

This method is useful for evaluating the partial area under the curve.

Gini index

The Gini index is defined as

$$\mathrm{Gini} = 2\mathrm{AUC} - 1. \tag{4.11}$$

This index is widely used in finance, *e.g.*, for credit scoring models assessment. It is only the transformation of AUC to the interval $[-1,1]$. For further details see, *e.g.*, Xu (2003); Anderson (2007); Thomas (2009).

Remark 4.4. *(Interpretation)* The Gini index takes values between -1 and 1. The value 1 corresponds to the perfectly accurate model (Fig. 4.4), the value 0 corresponds to the random model (Fig. 4.5), which is not usable for separation of groups $\mathcal{G}_0$ and $\mathcal{G}_1$, and negative values imply both groups should be changed with each other.

We would like to mention also another point of view of this index. The definition can be rewritten as

$$\mathrm{Gini} = \frac{\mathrm{AUC} - 1/2}{1/2}.$$

This formula represents the ratio of the area between the ROC curve for an actual model and the ROC for the random model (the diagonal) over the area between the ROC for the perfectly accurate model and the ROC for the random model. This idea will be useful in further considerations.

Partial area under the ROC curve

The AUC is the average performance over the entire range of possible sensitivities and specificities. This fact may cause some problems:

- two different curves can provide the same area
- not all regions of the ROC curve have the equal importance
- relevant sensitivities or specificities are often somewhere away from the ends of the ROC curve

To overcome these problems the Partial Area Under Curve (PAUC). The choice of the appropriate ranges depends on specific data and their properties (see Zhou *et al.* (2002)).

4.3.2   *Maximum improvement of sensitivity over chance diagonal (MIS)*

The MIS is the maximum difference in observed sensitivity and sensitivity at chance diagonal over all values of specificity. The corresponding $(1 - Sp)$ is denoted by p_{MIS}, *i.e.*, the maximum improvement of sensitivity.

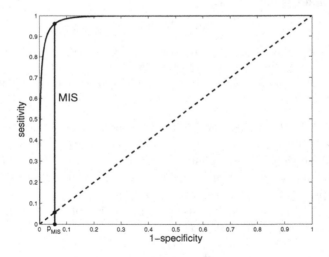

Fig. 4.7   Maximum improvement of sensitivity, MIS $= R(p_{\mathrm{MIS}}) - p_{\mathrm{MIS}}$.

This index can also be interpreted as follows. Assume $R(p)$ is concave. The point p_{MIS} is defined as an argument of maximum of the function $Q(p) = R(p) - p$, *i.e.*, zero of $Q'(p)$:

$$Q'(p) = R'(p) - 1, \quad R'(p) = \frac{f_1(c)}{f_0(c)}, \quad c = F_0^{-1}(1 - p).$$

Set $Q'(p) = 0$ and thus

$$\frac{f_1(\theta)}{f_0(\theta)} = 1 \Rightarrow f_1(\theta) = f_0(\theta),$$

$$\theta = F_0^{-1}(1 - p_{\text{MIS}}), \quad p_{\text{MIS}} = 1 - F_0^{-1}(\theta).$$

$R''(p) < 0 \Rightarrow p_{\text{MIS}}$ realizes the maximum of $Q(p) = R(p) - p$, *i.e.*, p_{MIS} is such a point of the ROC curve where a tangent slope is equal to one.

4.4 Other indices of reliability assessment

The objective of this section is to give an overview of other discrimination measures used to determine the quality of models at separating in a binary classification system. Especially, we will follow the financial sphere, where the discrimination power of scoring models is evaluated. However, most of all mentioned indices has wide application in many other areas, where models with binary output are used, like biology, medicine, engineering and so on.

References on this topic are quite extensive, see, *e.g.*, Siddiqi (2006); Anderson (2007); Thomas (2009). We summarize the most important quality measures and give some alternatives to them. All of the mentioned indices are based on the density or the distribution function, therefore one can suggest the technique of kernel smoothing for estimation. More detailed studies about all mentioned indices can be found, *e.g.*, in Koláček and Řezáč (2010); Řezáč and Koláček (2010).

4.4.1 *Cumulative Lift*

Let us consider the model with binary output defined in Sec. 4.1. One of possible indicators of the quality of the model can be the *cumulative Lift*.

Definition 4.2. Let X be one-dimensional absolute continuous random variable with the density $f(x) = p_0 f_0(x) + p_1 f_1(x)$, $p_0 > 0$, $p_1 > 0$, $p_0 + p_1 = 1$ and its related distribution function $F(x)$. For $c \in \mathbb{R}$, we define

$$\text{Lift}(c) = \frac{P(X \le c \mid D = 0)}{P(X \le c)} = \frac{F_0(c)}{F(c)}.$$

The function is called the *Lift*.

Remark 4.5. *(Interpretation)*
The ratio says how many times, at a given cutoff point c, the model is better than the random selection (random model). More precisely, the ratio indicates the proportion of subjects with a score X less than c from $\mathcal{G}_0$ to the proportion of these subjects in the general population.

Let us note the interesting fact that contrary to the measures based on ROC curve, the Lift formula depends on p_0, too, *i.e.*, the sizes of both groups are taken into account.

By using the notation $u = F(c)$ we arrive at the *cumulative Lift*

$$\text{QLift}(u) = \frac{F_0(F^{-1}(u))}{u}, \quad u \in [0, 1], \tag{4.12}$$

which is commonly used in real applications.

Figures 4.8 and 4.9 illustrate the cumulative Lift for two extreme models – the random model and the perfectly accurate model with $p_0 = 0.12$.

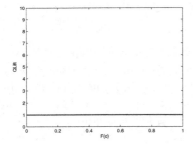

Fig. 4.8 Cumulative Lift for the random model.

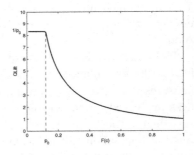

Fig. 4.9 Cumulative Lift for the perfectly accurate model.

For the case of the perfectly accurate model, the QLift takes the form

$$\text{QLift}_{PA}(u) = \begin{cases} 1/p_0, & 0 \leq u \leq p_0 \\ 1/u, & p_0 < u \leq 1. \end{cases} \tag{4.13}$$

In practical computation, the distribution functions F_0 and F in (4.12) are replaced by their empirical or kernel estimates. For more see, *e.g.*, Coppock (2002); Řezáč and Řezáč (2011).

Example 4.3. Figure 4.10 presents the cumulative Lift for the densities defined in Example 4.2 with $p_0 = 0.12$. In this case, the final curve is close to the cumulative Lift for the perfectly accurate model. The similar result appeared for the ROC curve, see Figure 4.6.

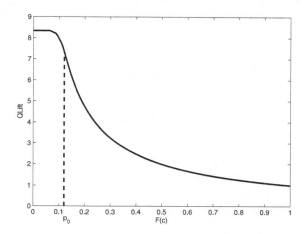

Fig. 4.10   Cumulative Lift for the model from Example 4.2.

4.4.2   *Lift Ratio*

As an analogy to the Gini index, we can choose a similar approach to derive the *Lift Ratio* (LR) index for the cumulative Lift

$$
\text{LR} = \frac{\int\limits_{0}^{1} \text{QLift}(u)\,du - 1}{\int\limits_{0}^{1} \text{QLift}_{PA}(u)\,du - 1}.
\tag{4.14}
$$

It is obvious that LR is a global measure of the model quality and that it takes values from 0 to 1. The value 0 corresponds to the random model, the value 1 matches the perfectly accurate model. An important feature is that the Lift Ratio allows us to fairly compare two models developed on different data samples, which is not possible with Lift.

Remark 4.6. The Lift Ratio for the cumulative Lift function from Example 4.3 takes the value LR $= 0.96$.

4.4.3 *Integrated Relative Lift*

The Lift Ratio compares areas under the Lift curve for the actual and the perfectly accurate models, but the next concept is focused on the comparison of Lift functions themselves. We define the *Relative Lift* function by

$$\mathrm{RLift}(u) = \frac{\mathrm{QLift}(u)}{\mathrm{QLift}_{PA}(u)}, \quad u \in (0,1].$$

Examples of this function are presented in Fig. 4.11 and Fig. 4.12 for two extreme models – the random model and the perfectly accurate model.

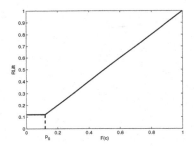

Fig. 4.11   Relative Lift for the random model.

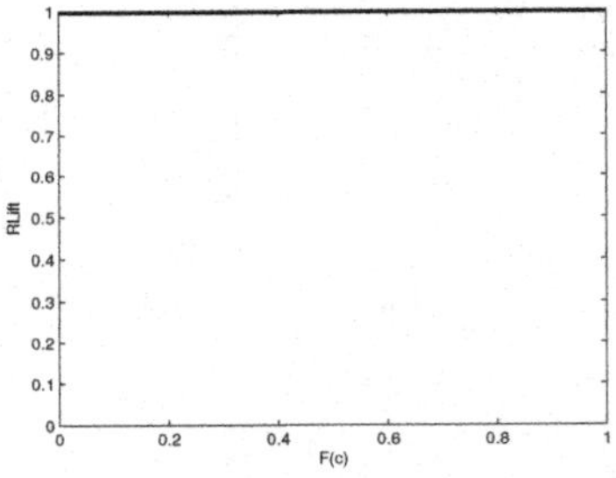

Fig. 4.12   Relative Lift for the perfectly accurate model.

Example 4.4. Figure 4.13 illustrates the relative Lift for the densities defined in Example 4.2 with $p_0 = 0.12$. Generally, the range of definition of RLift is $(0,1]$. The curve starts at the point $[u_{min}, p_0\mathrm{QLift}(u_{min})]$, where u_{min} is a positive number near to zero. It falls down to a local minimum at the point $[p_0, p_0\mathrm{QLift}(p_0)]$ and then it rises up to the point $[1,1]$. It is obvious that the graph of the Relative Lift function for a better model is closer to the top line which represents the function for the perfectly accurate model.

It is natural to extend our considerations to the area under the Relative Lift function. We define the *Integrated Relative Lift* (IRL) by

$$\mathrm{IRL} = \int_0^1 \mathrm{RLift}(u)\,du. \tag{4.15}$$

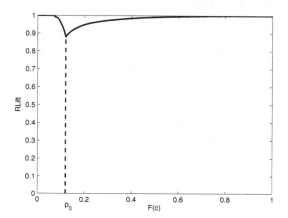

Fig. 4.13   Relative Lift for the model from Example 4.2.

It takes values from $0.5 + p_0^2/2$ (for the random model) to 1 (for the perfectly accurate model).

Remark 4.7. The IRL for the relative Lift function from Example 4.4 takes the value IRL $= 0.98$.

4.4.4   *Information Value*

The *Information Value* is related to the measures of entropy that appear in information theory and it is defined, *e.g.*, in Hand and Henley (1997); Thomas (2009).

Definition 4.3. Let the assumptions of Definition 4.2 be fulfilled and let the densities f_0, f_1 be continuous on $\mathbb{R}$. The *Information Value* is defined as

$$I_{val} = \int_{-\infty}^{\infty} f_{IV}(x)dx, \tag{4.16}$$

where

$$f_{IV}(x) = (f_1(x) - f_0(x)) \ln\left(\frac{f_1(x)}{f_0(x)}\right).$$

In statistics, this measure is sometimes called a *divergence*. The divergence is a continuous analogue to the information value. It was originally suggested by Kullback and Leibler (1951) as a way of measuring the relative distance between a "true" probability distribution and another one obtained from a model. In our context, it seeks to identify how f_0 and f_1 are different. Large values of I_{val} arise from large differences between f_0 and f_1 and thus correspond to more useful characteristics in differentiating $\mathcal{G}_0$ from $\mathcal{G}_1$.

Example 4.5. Figure 4.14 illustrates the function f_{IV} for the densities defined in Example 4.2. The related Information Value equals 12.25 in this case.

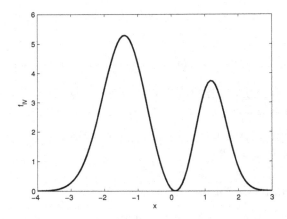

Fig. 4.14   Information Value for the model from Example 4.2.

In practice, it is useful to examine f_{IV} (y-axis) versus the distribution function F for score of all subjects (x-axis). This is illustrated in Fig. 4.15. The x-axis represents the so-called "rejection scale". It means, that we can see the value of f_{IV} for each percentile of rejected subjects (subjects determined as $\mathcal{G}_0$).

The Information Value is estimated empirically in practice. However, many computational problems arise here. The densities f_0 and f_1 are replaced by the relevant histograms. Obviously there exist some bins with zero observations of f_0 or f_1. Thus we get the logarithm of zero or dividing by zero in the estimation of (4.16) and a remedy is needed. For some approaches to the solution see, *e.g.*, Řezáč and Koláček (2011). With respect

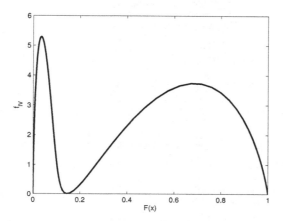

Fig. 4.15 f_{IV} vs. the rejection scale.

to these problems the kernel estimates of mentioned densities seem to be a good alternative (Koláček and Řezáč (2010)).

4.4.5 KR index

Let $c \in \mathbb{R}$ be a cutoff point value. Let us consider the confusion matrix given in Table 4.2 and rewrite it to the form given in Table 4.4. Symbols $n_{.j}$ and $n_{j.}$ $(j = 1, 2)$ denote column or row sums in the matrix, respectively.

Table 4.4   Confusion matrix

	Negative test $\mathcal{T} = 0$	Positive test $\mathcal{T} = 1$	Total
$\mathcal{G}_0$ $(D = 0)$	$P(X \leq c \mid D = 0)P(D = 0)$	$P(X > c \mid D = 0)P(D = 0)$	$n_{1.}$
$\mathcal{G}_1$ $(D = 1)$	$P(X \leq c \mid D = 1)P(D = 1)$	$P(X > c \mid D = 1)P(D = 1)$	$n_{2.}$
Total	$n_{.1}$	$n_{.2}$	

We can express the probabilities in the table by the distribution functions F_0, F_1. Thus we obtain Table 4.5.

Table 4.5   Confusion matrix

	Negative test $\mathcal{T} = 0$	Positive test $\mathcal{T} = 1$	Total
$\mathcal{G}_0$ $(D = 0)$	$F_0(c)p_0$	$(1 - F_0(c))p_0$	$n_1.$
$\mathcal{G}_1$ $(D = 1)$	$F_1(c)(1 - p_0)$	$(1 - F_1(c))(1 - p_0)$	$n_2.$
Total	$n._1$	$n._2$	

By using Pearson's χ^2-test of independence for the confusion matrix we arrive at the function $\chi^2(c)$

$$
\begin{aligned}
\chi^2(c) &= \frac{(n_{11}n_{22} - n_{12}n_{21})^2}{n._1 n._2 n_1. n_2.} \\
&= \frac{(F_0(c) - F_1(c))^2}{(F_0(c) - F_1(c))^2 + 1/p_0 F_1(c)(1 - F_1(c)) + 1/(1 - p_0) F_0(c)(1 - F_0(c))}.
\end{aligned}
$$

The value $\chi^2(c)$ describes the power of dependence of both groups for the given score value c. The higher values mean the better discrimination power of the model. Thus it is natural to consider the index KR, which is a maximum of $\chi^2(c)$

$$
\text{KR} = \max_{c \in \mathbb{R}} \chi^2(c). \tag{4.17}
$$

Let us denote $c_{\text{KR}} = \arg\max_{c \in \mathbb{R}} \chi^2(c)$ and mention some properties of the function χ^2. It takes values between 0 and 1 on the real line and it tends to zero for limit values of the score. In the case of the random model, it is identically equal to zero, since both groups have the same probability. In the case of the perfectly accurate model, there exists a cutoff point c_{KR}, for which $\chi^2(c_{\text{KR}}) = 1$.

Example 4.6. Figure 4.16 illustrates the function χ^2 for the distributions defined in Example 4.2. The related KR index takes the value 0.79 for $c_{\text{KR}} = -0.1818$ in this case.

The KR index is a type of "generalization" of MIS index. It reflects the proportion of subjects from $\mathcal{G}_0$, so it gives more information about the actual model than the MIS index. By the maximization process we obtain also the value of c_{KR} where the maximum is realized. Obviously, it is very useful to know the value of the "optimal" cutoff point for the given model. See Koláček and Řezáč (2011) for more details about this measure.

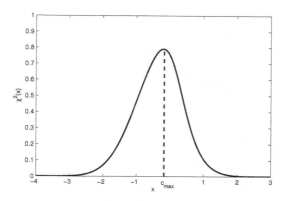

Fig. 4.16   Pearson's function for the model from Example 4.2.

4.5   Application to real data

4.5.1   *Head trauma data*

ROC curves are also useful for assessing the accuracy of variables to predict a patient's prognosis. In the first data set we consider the use of cerebrospinal fluid CK–BB (creative kinase–BB) isoenzyme measured within 24 hours of injury for predicting the outcome of severe head trauma (see Zhou *et al.* (2002)). We are interested in determining which patients have a poor outcome after suffering a severe head trauma. A sample of 60 subjects (group $\mathcal{G}_0$) admitted to a hospital with severe head trauma are considered, 19 (n_0) of whom eventually had moderate to full recovery and 41 (n_1) of whom (group $\mathcal{G}_1$) eventually had poor or no recovery. Thus the estimate $\hat{p}_0$ of the proportion p_0 equals $19/60 = 0.3167$.

Figure 4.17 shows the kernel estimate of the ROC curve with the removed boundary effects (see Sec. 3.4) for quartic kernel, $h_0 = 422.46$ and $h_1 = 733.59$ (solid line), the ROC curve obtained by binormal method (dashed line) and the empirical ROC curve (dotted line).

In Table 4.6 some summary indices are presented. The values of indices are very close to each other and similarly to the next examples, the final estimate of AUC is not too affected by the used method. Let us notice, that each index in the table has another interpretation and also the range. It is not easy to compare them and say the definite result about the prediction power of CK–BB isoenzyme. From the point of view of medicine the value

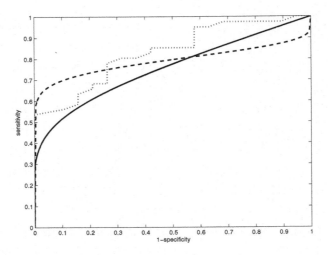

Fig. 4.17   Estimates of ROC curve for head trauma data.

of AUC seems to be sufficient and CK–BB isoenzyme could be considered as a good predictor of the outcome. On the other hand, the results of other indices are not so positive. These measures are mainly suitable for credit scoring models. They take less values in the context of these models and point to the moderate prediction power of the considered model.

Table 4.6   Summary indices for head trauma data

AUC by binormal model	0.8286
Empirical AUC	0.8286
AUC by comp. trap. rule	0.8286
Kernel estimate of AUC – method I	0.7188
Kernel estimate of AUC – method II	0.7940
Kernel estimate of AUC – method III	0.7573
MIS for kernel estimate	0.4155
c_{MIS}	0.1115
Gini	0.5802
QLift(p_0)	1.7730
LR	0.4293
IRL	0.7756
I_{val}	0.9912
KR	0.2095
c_{KR}	293.71

Finally, let us notice that the value $c_{KR} = 293.71$ sets an "optimal" (in the sense of KR index) cutoff value of CK–BB isoenzyme for predicting the outcome of severe head trauma.

4.5.2 *Pancreatic cancer data*

In the second data set the parametric and nonparametric methods were applied to real data set from Mayo Clinic, where sera from group of 51 patients with pancreatitis (group $\mathcal{G}_0$) and 90 patients with pancreatic cancer (group $\mathcal{G}_1$) were studied with a carbohydrate antigen assay (CA19-9) (see Zhou and Harezlak (2002) for details). We study a relative accuracy of the biomarker CA19-9 for 51 patients with condition and 90 patients without condition. Thus the estimate of the proportion p_0 is $\hat{p}_0 = 51/141 = 0.3617$.

The empirical ROC curve (dotted line), the parametric binormal method (dashed line) and the nonparametric kernel methods with the removed boundary effects (solid line) are presented in Fig. 4.18. The estimation methods were applied to the logarithmically transformed data.

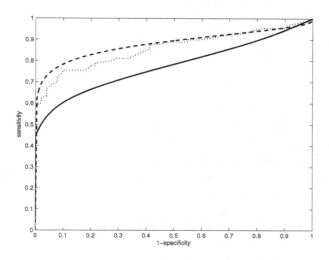

Fig. 4.18   Estimates of ROC curve for pancreatic cancer data.

Some summary indices for pancreatic cancer data are given in Table 4.7. Comparison the values of indices with the previous example leads to the conclusion that this model has a higher discrimination power.

Table 4.7   Summary indices for pancreatic cancer data

AUC by binormal model	0.8808
Empirical AUC	0.8614
AUC by comp. trap. rule	0.7772
Kernel estimate of AUC – method I	0.7766
Kernel estimate of AUC – method II	0.8458
Kernel estimate of AUC – method III	0.8379
MIS for kernel estimate	0.5053
p_{MIS}	0.0745
Gini	0.6918
$QLift(p_0)$	1.8826
LR	0.5071
IRL	0.8122
I_{val}	4.2027
KR	0.3544
c_{KR}	4.1638

Only the value of I_{val} differs essentially from the previous case. It is caused by the fact that the final estimate of I_{val} is very sensitive on the used method. Finally, let us notice that the value $c_{KR} = 4.1638$ sets an "optimal" (in the sense of KR index) cutoff value of the biomarker CA19-9 for predicting the pancreatic cancer.

4.5.3   *Consumer loans data*

In the last example, we demonstrate the use of mentioned indices in the financial area. The methods were applied to a data set from Home Credit a.s., where some (not specified) scoring function was used for predicting the likelihood of repayment of their clients. The company developed this scoring model to determine which clients are able to repay their loans.

The test data set consists of score values of 2 327 clients, 2 030 have repaid their loans (group $\mathcal{G}_1$) and 297 had problems with payments or did not pay (group $\mathcal{G}_0$). Thus the estimate $\hat{p}_0$ equals $297/2 327 = 0.1276$. We use all mentioned indices to assess the discrimination power of the given scoring function.

The comparison of parametric binormal method (dashed line) and nonparametric kernel methods with removed boundary effects (solid line) can are presented in Fig. 4.19. The empirical estimate was identical with the kernel estimate by reason of quite a lot of data.

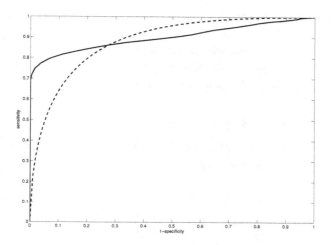

Fig. 4.19 Estimates of ROC curve for consumer loans data.

Table 4.8 summarizes all computed indices for consumer loans data. We note that they take the highest values in comparison with two previous examples. Thus we can conclude this model has the best performance.

Table 4.8 Summary indices for consumer loans data

AUC by binormal model	0.8810
Empirical AUC	0.9017
AUC by comp. trap. rule	0.8876
Kernel estimate of AUC – method I	0.8916
MIS for kernel estimate	0.5965
p_{MIS}	0.2121
Gini	0.8063
$\mathrm{QLift}(p_0)$	3.3342
LR	0.6120
IRL	0.8804
$\mathrm{I_{val}}$	8.5479
KR	0.3040
c_{KR}	0.9794

The value $c_{\mathrm{KR}} = 0.9794$ sets an "optimal" (in the sense of KR index) cutoff value of the client's score for predicting the ability to repay his loan.

4.6 Use of MATLAB toolbox

The toolbox can be downloaded from the web page
`http://www.math.muni.cz/english/science-and-research/`
`developed-software/232-matlab-toolbox.html.`

4.6.1 *Running the program*

The *Start menu* (Fig. 4.20) for kernel estimation of quality indices is called up by the command `ksquality`.

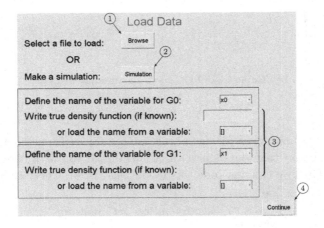

Fig. 4.20 Start menu.

You can skip this menu by typing input data as an argument `ksquality(x0,x1)`, where the vectors `x0` and `x1` are score results for two groups $\mathcal{G}_0$ and $\mathcal{G}_1$. If we know also their densities f_0 and f_1 (for example for simulated data), we can set them as the next arguments. For more see `help ksquality`. After the execution of this command the window in Fig. 4.24 is called up directly.

4.6.2 *Start menu*

In the *Start menu*, you have several possibilities how to define input data. You can load it from a file (button ①) or simulate data (button ②). In the fields ③ you can list your variables in the current workspace to define input data. If your workspace is empty, these fields are nonactive.

If you know the true densities for both groups, you can write them to text fields or load them from selected variables. If you need to simulate values for a model, press button ②. Then the menu for simulation (Fig. 4.21) is called up.

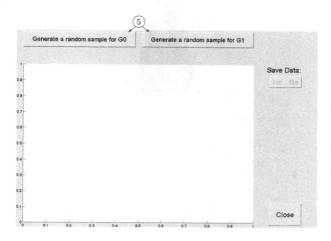

Fig. 4.21   Simulation menu – start.

4.6.3   *Simulation menu*

In the *Simulation menu*, first it is necessary to generate random samples for both groups by buttons ⑤. Either of these buttons calls up the *Data generating menu* (Fig. 4.22). In this menu, you can set the type of distribution and the size of a sample (fields ⑥). In the figure, the histogram for the generated sample is illustrated. In the bottom part of the menu, you can change parameters of a distribution and make a new sample by ⑦. If you have done the sample generating, you need to export your data. The button ⑧ calls up the menu where you specify the name of variable and confirm by pressing "OK". Then the *Data generating menu* will be closed and you will be returned to the *Simulation menu*. After data generating for both groups it looks like Fig. 4.23. In the figure, the histograms of generated samples for both groups are illustrated. The cyan color represents data for $\mathcal{G}_0$ and the red color is for $\mathcal{G}_1$. In this stage you can save data to variables or as a file by using buttons ⑨. If you have done the simu-

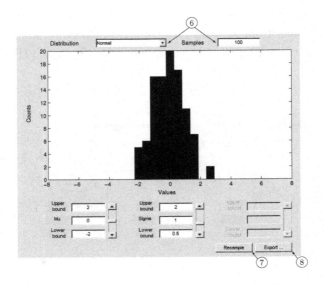

Fig. 4.22   Data generating menu.

lation, press button ⑩. The *Simulation menu* will be closed and you will
be returned to the *Start menu* (Fig. 4.20). In this menu, you can redefine
the input data. If you want to continue, press button ④. The menu will
be closed and the *Basic menu* (see the next paragraph) will be called up.

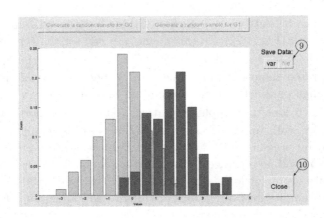

Fig. 4.23   Simulation menu.

4.6.4 *The final estimation*

This menu (Fig. 4.24) was called up from the *Start menu* or directly from the command line (see `help ksquality`). At the start of this application, you can see some color symbols in the figure. The blue crosses represent the score values $X_{01}, \ldots, X_{0n_0}$ for the first group $\mathcal{G}_0$, the red circles illustrate the score values $X_{11}, \ldots, X_{1n_1}$ for the second group $\mathcal{G}_1$.

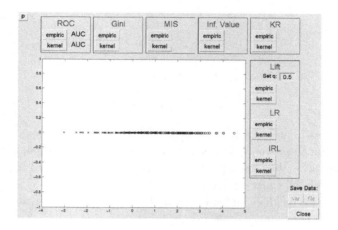

Fig. 4.24   Basic menu.

In the menu, there are six fields for computation of indices for reliability assessment. Each field contains two buttons for estimation. The "empiric" button represents the empirical estimate of the related index and the "kernel" button stands for the estimate obtained by kernel smoothing. For all cases the Epanechnikov kernel is used and the optimal smoothing parameter is estimated by the method of maximal smoothing principle. To avoid the boundary effects in estimation the reflection method is applied. For more details see Chapter 2 and 3.

Let us describe all fields more properly:

- "ROC" – in the first field you obtain the ROC curve for the actual model. At the right hand side of the used button the value of AUC (defined by (4.8)) is written. The actual curve is plotted in the figure, see Fig. 4.25.
- "Gini" – this field stands for the estimation of the Gini index defined in (4.11).

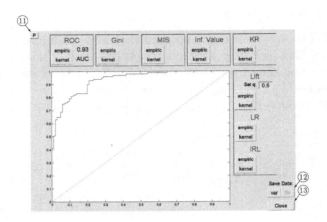

Fig. 4.25   Basic menu.

- "MIS" – estimates the Kolmogorov – Smirnov statistics denoted as MIS (Maximum Improvement of Sensitivity over chance diagonal, see §4.3.2).
- "Inf. Value" – the estimate of the Information Value (4.16) is computed and the related function f_{IV} is plotted.
- "KR" – this field is for estimating the KR index (4.17) and plotting the function χ^2.
- "Lift" – set the value q (from interval $(0,1)$) first and then estimate the cumulative Lift function (4.12) at this point, the estimate is plotted in the current figure. The terms "LR" and "IRL" stand for the estimation of the Lift Ratio (4.14) and the Integrated Relative Lift (4.15), respectively.

If you want to show only the actual curve, use button ⑪. You can also save data to variables and then as a file by using buttons ⑫. Button ⑬ ends the application.

Chapter 5

Kernel estimation of a hazard function

In recent years considerable attention has been paid to methods for analyzing data on events observed over time and to the study of factors associated with occurrence rate for these events. The problem of analyzing such data sets arises in a number of applied fields, such as medicine, biology, public health, epidemiology, technical sciences *etc.* Such data are generically referred to as survival data. A special feature of survival data which renders standard methods inappropriate is they contain censored observations.

Let T be a time until some specified event. This event may be death, the appearance of the tumor, equipment breakdown, but also remission after some treatment. We assume that T is a nonnegative random variable with absolute continuous distribution function F.

There are two functions characterizing the distribution of T, namely the survival function, which is the probability of an individual surviving to the time x, and the hazard function, which represents the instantaneous death rate for an individual surviving to time x. Parametric models for estimating these functions have been widely used. These models are chosen, not only because of popularity among researches, but also because they offer insight into the nature of the various parameters. Some of these important models include *e.g.*, the exponential, Weibull, gamma, lognormal, Gompertz, Pareto distribution *etc.* – see *e.g.*, Table 2.2 in Klein and Moeschberger (2003), Hougaard (2001) for detailed discussion or Collet (1997); Thernau and Grambsch (2001).

In order to avoid assuming that some parametric model is correct, it is common to use nonparametric methods. The simplest nonparametric estimation of the survival function for complete observations is an empirical survival function. Kaplan and Meier (1958) suggested an extension to censored data. Estimation of the hazard function has received relatively less attention than the survival function. Watson and Leadbetter (1964) and Tanner and Wong (1983, 1984) studied parametric and nonparametric estimations of the hazard function. Patil *et al.* (1994) considered two types of nonparametric hazard function estimators and derived their asymptotic properties. Kernel-type estimators were also studied in Ramlau-Hansen (1983); Yandell (1983); Mielniczuk (1986); Müller and Wang (1990a,b, 1994); Uzunogullari and Wang (1992); Patil (1993a,b); Nielsen and Linton (1995); Youndjé *et al.* (1996); Jiang and Marron (2003). In paper by Kim *et al.* (2005) the simple nonparametric estimator of hazard function was proposed by using the Kaplan–Meier estimator.

5.1 Basic definition

The survival time or lifetime, *i.e.*, the random variable T, is interpreted to be the time from the beginning of follow-up to the death or to any event under consideration. Let us denote the distribution function of T by F, *i.e.*, $F(x) = P(T < x)$. A survival process can be characterized by the *survival function* $\bar{F}$

$$\bar{F}(x) = P(T \geq x) = 1 - F(x). \tag{5.1}$$

The survival function $\bar{F}(x)$ is defined as the probability that the survival time is greater or equal to x. The hazard function is the probability that an individual dies at the time x, conditional on he or she having survived to that time. If the life distribution F has a density f, for $\bar{F}(x) > 0$, the *hazard function* is defined by

$$\lambda(x) = \frac{f(x)}{\bar{F}(x)} \tag{5.2}$$

and the *cumulative hazard function* as

$$H(x) = -\log \bar{F}(x). \tag{5.3}$$

We consider the random censorship model, where data are censored from the right. This type of censorship is often met in many applications, especially in clinical research or in the life testing of complex technical systems.

We only remind that the survival time of the individual is right censored if the end-point of interest has not been observed for that individual. The right censored survival time is then less than the actual, but unknown, survival time.

Let $T_1, T_2, \ldots, T_n$ be i.i.d. (independent and identically distributed) survival times with the distribution function F. Let $C_1, C_2, \ldots, C_n$ be i.i.d. censoring times with the distribution function G which are usually assumed to be independent from the survival times.

In the random censorship model we observe the pairs

$$(X_i, \delta_i), i = 1, 2, \ldots, n, \text{ where } X_i = \min(T_i, C_i)$$

and $\delta_i = I_{\{X_i = T_i\}}$ indicates whether the observation is censored or not. It follows that $\{X_i\}$ are i.i.d. with survival function $\bar{L}$

$$\bar{L}(x) = \bar{F}(x)\bar{G}(x)$$

($\bar{F}$ and $\bar{G}$ are survival functions for $\{T_i\}$ and $\{C_i\}$, respectively).

In 1958 Kaplan and Meier proposed the estimate of the survival function $\bar{F}$:

$$\widehat{\bar{F}}(x) = \prod_{\{j : X_{(j)} < x\}} \left(\frac{n-j}{n-j+1} \right)^{\delta_{(j)}} \tag{5.4}$$

where $X_{(j)}$ denotes the j-th order statistics of $X_1, X_2, \ldots, X_n$ and $\delta_{(j)}$ the corresponding indicator of the censoring status. $\widehat{\bar{F}}(x)$ is a step function and it is right continuous. The estimator is also called the product-limit estimator because of its derivation, as a limit, when the time is split into intervals and the interval length goes to 0.

We also remind the modified empirical survival function of observation times $\bar{L}_n = 1 - L_n$, where

$$L_n(x) = \frac{1}{n+1} \sum_{i=1}^{n} I_{\{X_i \le x\}}. \tag{5.5}$$

Nelson (1972) proposed to estimate the cumulative hazard function H by

$$\mathcal{H}_n(x) = \sum_{X_{(i)} \le x} \frac{\delta_{(i)}}{n-i+1} \tag{5.6}$$

We focus on kernel-type estimates introduced by Tanner and Wong (1983); Müller and Wang (1990b) and Jiang and Marron (2003).

The kernel estimate of the ν-th derivative of the hazard function λ is the following convolution of the kernel K with the Nelson estimator $\mathcal{H}_n$:

$$\widehat{\lambda}^{(\nu)}(x, h) = \frac{1}{h^{\nu+1}} \int K^{(\nu)} \left(\frac{x-u}{h} \right) d\mathcal{H}_n(u) \tag{5.7}$$

$$= \frac{1}{h^{\nu+1}} \sum_{i=1}^{n} K^{(\nu)} \left(\frac{x - X_{(i)}}{h} \right) \frac{\delta_{(i)}}{n-i+1}, \quad K^{(\nu)} \in S_{\nu,k}^0.$$

5.2   Statistical properties of the estimate

The properties of this estimate have been investigated by Müller and Wang (1990a) and the following theorem has been proved:

Theorem 5.1.

- *Let $[0, S], S > 0$, be such an interval for which $L(S) < 1$,*
- *$\lambda \in C^{k_0}[0, S], k_0 \geq 2, k \leq k_0$,*
- *$K^{(\nu)} \in S^0_{\nu,k}$,*
- *Let $h = h(n)$ be bandwidths satisfying*

$$\lim_{n \to \infty} h = 0, \quad \lim_{n \to \infty} h^{2\nu+1}n = \infty$$

$$\lim_{n \to \infty} nh^{k+1}(\log n)^{-1} = \infty, \quad \lim_{n \to \infty} nh(\log n)^{-2} = \infty.$$

Then, the Mean Square Error at the point $x \in (0, S]$ can be expressed by the formula

$$\mathrm{MSE}\{\widehat{\lambda}^{(\nu)}(x,h)\} = \underbrace{\left[h^{k-\nu}\lambda^{(k)}(x)\left\{\frac{(-1)^k\beta_k}{k!} + o(1)\right\}\right]^2}_{\mathrm{bias}^2\{\widehat{\lambda}^{(\nu)}(x,h)\}}$$

$$+ \underbrace{\frac{1}{nh^{2\nu+1}}\left\{\frac{\lambda(x)V(K^{(\nu)})}{\bar{\bar{L}}(x)} + o(1)\right\}}_{\mathrm{var}\{\widehat{\lambda}^{(\nu)}(x,h)\}}. \tag{5.8}$$

The global quality of this estimate can be described by means of the Mean Integrated Square Error (MISE$\{\widehat{\lambda}^{(\nu)}(\cdot, h)\}$).

Now, we focus on the Asymptotic Mean Integrated Square Error. Evidently, AMISE$\{\widehat{\lambda}^{(\nu)}(\cdot, h)\}$ takes the form

$$\mathrm{AMISE}\{\widehat{\lambda}^{(\nu)}(\cdot, h)\} = h^{2(k-\nu)}\beta_k^2 D_k + \frac{V(K^{(\nu)})\mathcal{L}}{nh^{2\nu+1}} \tag{5.9}$$

for

$$\beta_k = \beta_k(K^{(\nu)}) = \int_{-1}^{1} x^k K^{(\nu)}(x)dx, \quad D_k = \int_{0}^{S}\left(\frac{\lambda^{(k)}(x)}{k!}\right)^2 dx$$

and

$$\mathcal{L} = \int_{0}^{S}\frac{\lambda(x)}{\bar{\bar{L}}(x)}dx.$$

Then, the asymptotically optimal bandwidth $h_{opt,\nu,k}$ minimizing AMISE$\{\widehat{\lambda}^{(\nu)}(\cdot, h)\}$ over the set H_n of acceptable bandwidths (see Sec. 5.3.4) with respect to h is given by

$$h_{opt,\nu,k}^{2k+1} = \frac{\mathcal{L}(2\nu + 1)}{2n(k - \nu)D_k}\gamma_{\nu,k}^{2k+1}. \tag{5.10}$$

The formula (5.10) provides a simple insight into an "optimal" bandwidth. But an obvious problem of finding $h_{opt,\nu,k}$ similarly as in the previous kernel estimates is that the optimal bandwidth depends on $\mathcal{L}$ and D_k.

As in density estimations this formula offers a very useful tool for the calculation of the optimal bandwidth for $\hat{\lambda}^{(\nu)}$ by means of bandwidth for $\hat{\lambda}^{(0)}$ and $\hat{\lambda}^{(1)}$ (see formulas (2.33), (2.34)). Further, continuing in a similar way as in the second chapter we arrive at the following expression (see (2.35), (2.36))

$$\text{AMISE}\left\{\widehat{\lambda}^{(\nu)}(\cdot, h_{opt,\nu,k})\right\} = \mathcal{L}\, T(K^{(\nu)})\frac{(2k + 1)\gamma_{\nu,k}^{2\nu+1}}{2n(k - \nu)h_{opt,\nu,k}^{2\nu+1}}. \tag{5.11}$$

This formula shows the effects of the kernel and its order as well as the bandwidth on the estimate. It looks as the formula (2.12) for kernel density estimates apart from the quantity $\mathcal{L}$.

The confidence intervals for the estimate $\widehat{\lambda}^{(\nu)}(\cdot, h_{opt,\nu,k})$ can be constructed. Their construction is described in the paper by Müller and Wang (1990b). The asymptotic $(1 - \alpha)$ confidence interval for $\lambda^{(\nu)}(x, h)$ is given by

$$\widehat{\lambda}^{(\nu)}(x, h) \pm \left\{\frac{\widehat{\lambda}(x, h)V(K)}{(1 - L_n(x))n\hat{h}^{2\nu+1}}\right\}^{1/2}\Phi^{-1}(1 - \alpha/2) \tag{5.12}$$

where Φ is the standard normal distribution function.

Remark 5.1. (*Choosing the shape of the kernel*)
The formula (5.11) suggests naturally looking for a kernel minimizing the functional $T(K)$. This problem has been investigated in Chapter 2 and the above-mentioned results can be applied for kernel estimates of the hazard function.

5.3 Choosing the bandwidth

Some of ideas of methods for density estimations can be transferred to the hazard function estimations. We only focus of problem of bandwidth selection for the hazard function itself, because in application we will deal with the second derivative of the hazard function and in such a case we can use formula (2.33) with $\nu = 2$.

5.3.1   *Cross-validation method*

Due to censoring modified cross-validation methods are applied for the estimate of the optimal bandwidth (see *e.g.*, Marron and Padgett (1987); Uzunogullari and Wang (1992); Nielsen and Linton (1995)). The main idea of this method is to use the Integrated Square Error in the form

$$
\begin{aligned}
\text{ISE}\{\widehat{\lambda}(\cdot, h)\} &= \int\limits_0^S \left(\widehat{\lambda}(x, h) - \lambda(x)\right)^2 dx \\
&= \int\limits_0^S \widehat{\lambda}^2(x, h)dx - 2\int\limits_0^S \frac{\widehat{\lambda}(x,h)}{1-F(x)}f(x)dx + \int\limits_0^S \lambda^2(x)dx .
\end{aligned}
\tag{5.13}
$$

The third term does not depend on h and the unbiased estimate of the integral in the second term (see Patil (1993a)) takes the form

$$
\frac{1}{n}\sum_{i=1}^n \frac{\widehat{\lambda}_{-i}(X_i, h)}{1 - L_n(X_i)}\delta_i,
$$

where $\widehat{\lambda}_{-i}(X_i, h)$ is the estimate of the hazard function in X_i omitting this point. We consider the cross-validation function

$$
\text{CV}(h) = \int\limits_0^S \widehat{\lambda}^2(x, h)dx - \frac{2}{n}\sum_{i=1}^n \frac{\widehat{\lambda}_{-i}(X_i, h)}{1 - L_n(X_i)}\delta_i,
\tag{5.14}
$$

and cross-validation bandwidth selection is denoted by $\widehat{h}_{\text{CV}}$:

$$
\widehat{h}_{\text{CV}} = \arg\min_{h \in H_n} \text{CV}(h),
$$

H_n denotes the set of acceptable bandwidths.

5.3.2   *Maximum likelihood method*

The modified likelihood method was proposed in the paper by Tanner and Wong (1984). The idea of the method is to choose h maximizing the modified likelihood function

$$
\text{ML}(h) = \prod_{i=1}^n \widehat{\lambda}_{-i}^{\delta_i}(X_i, h)\bar{F}_{-i}(X_i)
$$

where $\widehat{\lambda}_{-i}(X_i, h)$ is the same as in the cross-validation method and $\bar{F}_{-i}(x) = \exp\left(-\int_0^x \widehat{\lambda}_{-i}(t, h)dt\right)$. The selected bandwidth is denoted by

$$h_{\mathrm{ML}} = \arg\max_{h \in H_n} \mathrm{ML}(h).$$

Other methods for an estimate of the bandwidth can be also found *e.g.*, in papers by Sarda and Vieu (1991); Patil (1993a,b); Patil *et al.* (1994); González-Manteiga *et al.* (1996).

5.3.3 Iterative method

We again apply the idea of the iterative method described in Chap. 2. It is easy to prove the statement (Horová and Zelinka (2007b); Horová *et al.* (2006)):

Lemma 5.1.

$$\mathrm{AIV}\{\widehat{\lambda}(\cdot, h_{opt,0,k})\} - 2k\mathrm{AISB}\{\widehat{\lambda}(\cdot, h_{opt,0,k})\} = 0. \tag{5.15}$$

The estimate of AMISE is defined as

$$\widehat{\mathrm{AMISE}}\left\{\widehat{\lambda}(\cdot, h)\right\} = \int_0^S \widehat{\mathrm{var}}\{\widehat{\lambda}(x, h)\}dx + \int_0^S \widehat{\mathrm{bias}}^2\{\widehat{\lambda}(x, h)\}dx, \tag{5.16}$$

where $\widehat{\mathrm{var}}\{\widehat{\lambda}(x, h)\}$ and $\widehat{\mathrm{bias}}\{\widehat{\lambda}(x, h)\}$ are the estimates of variance and bias, respectively, given by

$$\widehat{\mathrm{var}}\left\{\widehat{\lambda}(x, h)\right\} = \frac{1}{nh} \int K^2(y) \frac{\widehat{\lambda}(x-hy, h)}{\bar{L}_n(x-hy)} dy,$$

$$\widehat{\mathrm{bias}}\left\{\widehat{\lambda}(x, h)\right\} = \int \widehat{\lambda}(x - hy, h)K(y)dy - \widehat{\lambda}(x, h), \tag{5.17}$$

where $K \in S_{0k}^0$ and $\bar{L}_n$ (see (5.5)) is the modified empirical survival function.

As in Chap. 2 $\hat{h}_{\mathrm{IT}}$ will stand for minimization of $\widehat{\mathrm{AMISE}}\left\{\widehat{\lambda}(\cdot, h)\right\}$

$$\hat{h}_{\mathrm{IT}} = \arg\min_{h \in H_n} \widehat{\mathrm{AMISE}}\left\{\widehat{\lambda}(\cdot, h)\right\}. \tag{5.18}$$

Taking into account the results of Müller and Wang (1990a,b) and our considerations given in Chap. 2 we arrive at the fact that in probability

$$\frac{\hat{h}_{\mathrm{IT}}}{h_{opt,0,k}} \to 1.$$

According to Lemma 5.1 minimization of $\widehat{\mathrm{AMISE}}\left\{\widehat{\lambda}(\cdot,h)\right\}$ is equivalent to solving equation:

$$\frac{1}{nh}\int\limits_0^S\int K^2(y)\frac{\widehat{\lambda}(x-hy,h)}{\bar{L}_n(x-hy)}\,dydx \quad -$$

$$-\ 2k\int\limits_0^S\left(\int\widehat{\lambda}(x-hy,h)K(y)dy-\widehat{\lambda}(x,h)\right)^2 dx=0,$$

which can be rewritten as

$$h=\frac{1}{2kn}\frac{\int\limits_0^S\int K^2(y)\frac{\widehat{\lambda}(x-hy,h)}{\bar{L}_n(x-hy)}\,dydx}{\int\limits_0^S\left(\int\widehat{\lambda}(x-hy,h)K(y)dy-\widehat{\lambda}(x,h)\right)^2 dx},\qquad(5.19)$$

and denoting the right hand side of this equation by g we obtain the non-linear equation

$$h=g(h),$$

and we are seeking for the fixed point of the function g. Thus, we can proceed as the same way as in Chap. 3 and use Steffensen's method for detecting the fixed point of the function g.

The evaluation of the function g looks rather complicated, but these integrals can be easily evaluated by suitable discretization; here the composite trapezoidal rule is recommended (see Horová and Zelinka (2007b)).

5.3.4 *Acceptable bandwidths*

Let us come back to the set H_n of acceptable bandwidths. The good choice of the initial approximation is very important for the above mentioned iterative process as well as for cross-validation and maximal likelihood methods (see also Horová and Zelinka (2007b)).

We will show how a kernel density estimate could be useful for this aim. Let us describe the motivation of our procedure. There is a natural question about the distribution of the time censor C. There are both theoretical and practical reasons for adopting the Koziol–Green model of random censorship (see Koziol and Green (1976)) under which it is assumed that there is a nonnegative constant ρ such that

$$\bar{F}(x)^\rho=\bar{G}(x),$$

$\rho = 0$ corresponds to the case without censoring.

Let l, f and g be densities of L, F and G, respectively. Then according to Hurt (1992)

$$l(x) = \frac{1}{p}\bar{F}(x)^p f(x), \tag{5.20}$$

$$p = \frac{1}{1+\rho}.$$

Let

$$\hat{l}(x,h) = \frac{1}{nh}\sum_{i=1}^{n} K\left(\frac{x-X_i}{h}\right) \tag{5.21}$$

be the kernel estimate of the density l and keep now $\bar{F}(x)$ as a known quantity. Then, with respect to (5.20)

$$\tilde{f}(x,h) = \frac{p\hat{l}(x,h)}{\bar{F}^p(x)}$$

and $\tilde{f}(x,h)$ is an estimate of f.

Consider now an alternative estimate $\tilde{\lambda}(\cdot, h)$ of λ:

$$\tilde{\lambda}(x,h) = \frac{\tilde{f}(x,h)}{\bar{F}(x)}. \tag{5.22}$$

Hence

$$\tilde{\lambda}(x,h) = \frac{p\hat{l}(x,h)}{\bar{F}^{1/p}(x)}. \tag{5.23}$$

Now it is easy to verify that the optimal bandwidth for $\hat{l}(\cdot, h)$ is also the optimal bandwidth for $\tilde{\lambda}(\cdot, h)$. The properties of the estimate $\tilde{\lambda}(\cdot, h)$ can be investigated in a similar way as those in the paper by Uzunogullari and Wang (1992). Let $\hat{h}^*_{opt,0,k}$ be an estimate of the optimal bandwidth for $\hat{l}(\cdot, h)$

$$h^*_{opt,0,k} = \arg\min_{h \in H_n} \text{AMISE}\left\{\hat{l}(\cdot, h)\right\}, \tag{5.24}$$

(for H_n see Chap. 2). Due to the above mentioned facts it is reasonable to take this value as a suitable initial approximation for the iterative process.

5.3.5   *Points of the most rapid change*

In the biomedical application the points θ of the most rapid change, *i.e.*, the points of the extreme of the first derivative of λ, are also of a great interest. These points can be detected as zeros of the estimated second derivatives. Thus, we will only concentrate on the estimate of $\lambda^{(0)}$ and $\lambda^{(2)}$. We focus on such points $\widehat{\theta}$, $\widehat{\lambda}^{(2)}(\widehat{\theta}, h) = 0$, where $\widehat{\lambda^{(2)}}(\cdot, h)$ changes its sign from $-$ to $+$, since only the local minima of $\widehat{\lambda}^{(1)}(\cdot, h)$ are important. It can be shown that $\widehat{\theta} \to \theta$ in probability (Müller and Wang (1990b)). The zero point $\widehat{\theta}$ of the estimate $\widehat{\lambda^{(2)}}(\cdot, h)$ are found by using some numerical method (*e.g.*, the secant method). It is possible that there exist several points of the change as we will see in Sec. 5.5.

     Confidence intervals for the point $\widehat{\theta}$ of the most rapid change are given by

$$\widehat{\theta} \pm \left[\frac{\widehat{\lambda}(\widehat{\theta}, h)V(K)}{(1 - L_n(\widehat{\theta}))\widehat{\lambda}^{(3)}(\widehat{\theta}, h)n\widehat{h}^5} \right]^{1/2} \Phi^{-1}(1 - \alpha/2),$$

where $\widehat{\lambda}^{(3)}(\widehat{\theta}, h)$ is approximated by divided difference of the second derivative (see Müller and Wang (1990b)).

5.4   Description of algorithm

According to our experience the kernel $K \in S_{0,4}^0$

$$K_{opt,0,4}(x) = \frac{15}{32}(x^2 - 1)(7x^2 - 3), \quad \gamma_{04} = 2.0165 \qquad (5.25)$$

is very convenient for the estimate of the hazard function.

     In connection with theoretical results the kernel $K^{(2)} \in S_{2,4}^0$

$$K_{opt,2,4}^{(2)}(x) = \frac{105}{16}(1 - x^2)(5x^2 - 1), \gamma_{24} = 1.3925 \qquad (5.26)$$

should be chosen for the estimate of $\lambda^{(2)}$.

Description of the algorithm:

Step 1. Estimate the density l with the kernel (5.25) and find the estimate of the optimal bandwidth $\widehat{h}_{opt,0,4}^*$ (see (5.24)).

Step 2. Put $\widehat{h}_{opt,0,4}^* = \widehat{h}_{opt,0,4}^{(0)}$ and use this value as the initial approximation for the iterative method which yields $\widehat{h}_{opt,0,4}$.

Step 3. Construct the estimate $\widehat{\lambda}^{(0)}$ with the kernel (5.25) and the bandwidth obtained in the second step.

Step 4. Compute the optimal bandwidth for the estimate $\widehat{\lambda}^{(2)}(\cdot, h)$ using the formula (2.33):

$$\hat{h}_{opt,2,4} = (10)^{1/9} \frac{\gamma_{24}}{\gamma_{04}} \hat{h}_{opt,0,4} = 0.8919 \hat{h}_{opt,0,4}.$$

Step 5. Get the kernel estimate of $\lambda^{(2)}$ with the kernel (5.26) and the bandwidth selected in the step 4.

Step 6. Detect the zeros of $\widehat{\lambda}^{(2)}(\cdot, h)$ and construct the confidence intervals.

Remark 5.2. When the estimate is near 0 or S then boundary effects can occur. This can lead to negative estimates of hazard functions near the endpoints. The same can happen if kernels of higher order are used in the interior. In such cases it may be reasonable to truncate $\widehat{\lambda}(\cdot, h)$ below at 0, *i.e.*, to consider $\widehat{\lambda}(x, h) = \max(\widehat{\lambda}(x, h), 0)$. The similar considerations can be made for the confidence intervals. The boundary effects can be avoided by using kernels with asymmetric supports (see *e.g.*, Müller and Wang (1990b, 1994) and also Chap. 2 and Chap. 3).

5.5 Application to real data

5.5.1 *Breast carcinoma data*

The first data set was kindly provided by the Masaryk Memorial Cancer Institut in Brno (Soumarová *et al.* (2002)).

The set of data involves 236 patients with breast carcinoma (BC). The study was carried out based on the records of women who had received the breast conservative surgical treatment and radiotherapy as well in the period 1983-1994. 47 patients died by the end of the study and 189 were thus censored. Number of deaths in particular years are seen in Fig. 5.1. The study was finished in the year 2000. Survival time is given in months from the surgical treatment to the end of the study.

In Fig. 5.2 the Kaplan–Meier estimate of the survival function $\bar{F}$ is presented and Fig. 5.3 shows the shape of the iterative function defined in (5.19). Figure 5.4 brings the estimate $\widehat{\lambda}^{(0)}(\cdot, h)$ constructed by the proposed procedure including the confidence intervals and Fig. 5.5 shows the estimate $\widehat{\lambda}^{(2)}(\cdot, h)$. Estimated points of the most rapid change $\widehat{\theta}_1, \widehat{\theta}_2$ are defined as zero of the estimated the second derivatives with sign changes from $-$ to $+$. The main change obviously occurs for $\widehat{\theta}_1 \doteq 51.39$ months

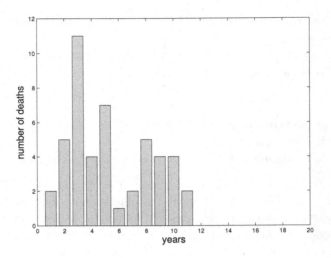

Fig. 5.1   Deaths in particular years for BC data.

while as the second change at $\widehat{\theta}_2 \doteq 128.87$ months. These figures indicated that patients run a high risk for approximated 50 months after surgical treatment. Then it is followed by a low risk and higher risk occurs again in the 100th month approximately.

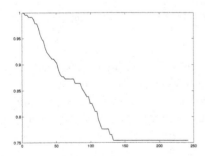

Fig. 5.2   Kaplan–Meier estimate of the survival function $\bar{F}$ for BC data.

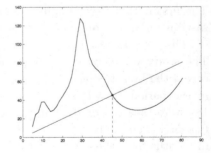

Fig. 5.3   Iterative function and the fixed point for BC data.

Basic characteristics and results concerned above mentioned data are summarized in Table 5.1.

Table 5.2 brings the sequence of iterations generated by the iterative method for tolerance $\varepsilon = 1.0 \times 10^{-6}$.

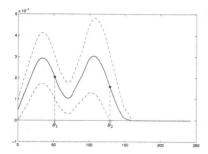

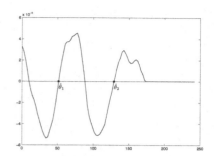

Fig. 5.4 Estimate of the hazard function $\widehat{\lambda}^{(0)}(\cdot, h)$ (solid line), the confidence intervals (dashed line) for BC data.

Fig. 5.5 Estimate of $\widehat{\lambda}^{(2)}(\cdot, h)$ for BC data.

Table 5.1 BC data.

number of patients	n	=	236
max. follow-up	S	=	220 months
number of deaths	n_d	=	47
percentage of deaths	p_d	=	19.9%
bandwidths	$\hat{h}_{opt,0,4}$	=	45.205
	$\hat{h}_{opt,2,4}$	=	40.319
most rapid change	$\hat{\theta}_1$	=	51.393
	$\hat{\theta}_2$	=	128.869

Table 5.2 Sequence of iterations.

j	0	1	2	3
$\hat{h}^{(j)}_{opt,0,k}$	22.526490	51.517280	47.411171	45.548046

j	4	5	6	7
$\hat{h}^{(j)}_{opt,0,k}$	45.222855	45.205395	45.205249	45.205249

5.5.2 *Cervix carcinoma data*

The second data set was also provided by the Masaryk Memorial Cancer Institut in Brno (Doleželová *et al.* (2008)). It is based on 51 patients with cervix carcinoma (CC). Survival time is given in months. It represents the number of months from the radiotherapy to the return of the disease.

Figure 5.6 presents number of events in separate months. In Figures 5.7 to 5.10, the same outputs as for the first data set are presented, *i.e.*, Kaplan–Meier estimate of the survival function, the iterative function and the esti-

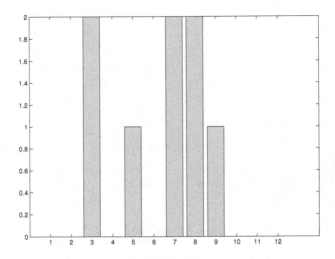

Fig. 5.6   Events in particular months for CC data.

mates of $\widehat{\lambda}^{(0)}(\cdot, h)$ and $\widehat{\lambda}^{(2)}(\cdot, h)$. The overview of the basic characteristics and results can be found in Table 5.3.

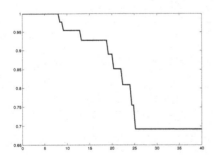

Fig. 5.7   Kaplan–Meier estimate of the survival function $\bar{F}$ for CC data.

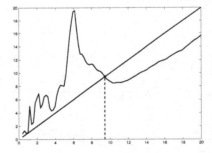

Fig. 5.8   Iterative function and the fixed point for CC data.

5.5.3   *Chronic lymphocytic leukaemia*

The third data set is based on record of 73 patients with chronic lymphocytic leukaemia (CLL) treated by Alemtuzumab (see Doubek *et al.* (2009)). Survival times are presented by the number of days from the date of diagnosis to the date of the last follow up or death.

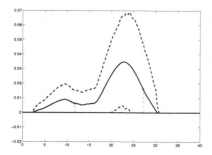

Fig. 5.9 Estimate of the hazard function $\widehat{\lambda}^{(0)}(\cdot, h)$ (solid line), the confidence intervals (dashed line) for CC data.

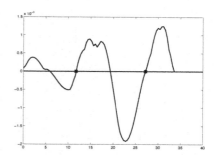

Fig. 5.10 Estimate of $\widehat{\lambda}^{(2)}(\cdot, h)$ for CC data.

Table 5.3   CC data.

number of patients	n	$=$	51
max. follow-up	S	$=$	36 months
number of deaths	n_d	$=$	8
percentage of deaths	p_d	$=$	15.7%
bandwidths	$\hat{h}_{opt,0,4}$	$=$	9.4341
	$\hat{h}_{opt,2,4}$	$=$	8.414
most rapid change	$\hat{\theta}_1$	$=$	11.759
	$\hat{\theta}_2$	$=$	27.210

Figure 5.11 presents number of deaths in separate years. Figures 5.12 to 5.15 and Table 5.4 are of the same meaning as for previous data sets.

Table 5.4   CLL data.

number of patients	n	$=$	73
max. follow-up	S	$=$	4982 days$\approx$13.5 years
number of deaths	n_d	$=$	33
percentage of deaths	p_d	$=$	45.2%
bandwidths	$\hat{h}_{opt,0,4}$	$=$	$1,708.0$
	$\hat{h}_{opt,2,4}$	$=$	$1,523.4$
most rapid change	$\hat{\theta}_1$	$=$	$1,225.1$
	$\hat{\theta}_2$	$=$	$3,025.4$
	$\hat{\theta}_3$	$=$	$4,917.1$

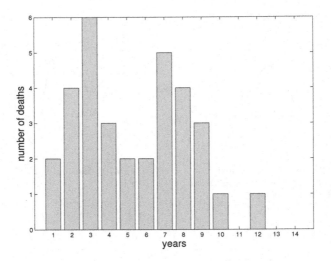

Fig. 5.11   Deaths in particular years for CLL data.

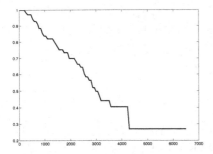

Fig. 5.12   Kaplan–Meier estimate of the survival function $\bar{F}$ for CLL data.

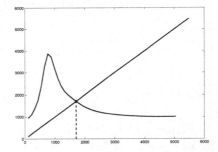

Fig. 5.13   Iterative function and the fixed point for CLL data.

5.5.4   *Bone marrow transplant*

The last data set has been taken from Klein and Moeschberger (2003). It consists of the 38 records of bone marrow transplanted patients with acute lymphoblastic leukemia (ALL). The patients were followed until death or the end of the study. The death rate (63.2%) is the highest from all the data sets.

Figure 5.16 presents number of events in separate years. The meaning of presented figures and table (Fig. 5.17 to Fig. 5.20 and Table 5.5) is the same as for the previous data sets.

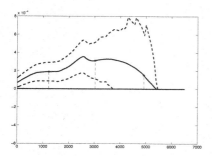

Fig. 5.14 Estimate of the hazard function $\widehat{\lambda}^{(0)}(\cdot, h)$ (solid line), the confidence intervals (dashed line) for CLL data.

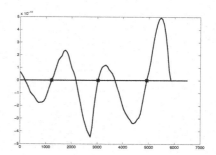

Fig. 5.15 Estimate of $\widehat{\lambda}^{(2)}(\cdot, h)$ for CLL data.

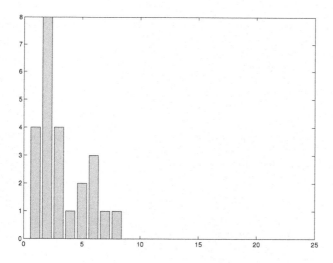

Fig. 5.16 Events in particular years for ALL data.

Remark 5.3. Papers Horová *et al.* (2007, 2009) are dealing with an estimate of an analytical form of the hazard function. They combine a kernel method with a deterministic approach based on cancer cell population dynamic (see also Kozusko and Bajzer (2003)).

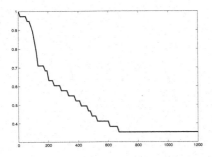

Fig. 5.17 Kaplan–Meier estimate of the survival function $\bar{F}$ for ALL data.

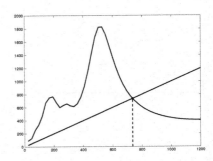

Fig. 5.18 Iterative function and the fixed point for ALL data.

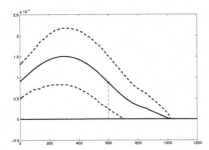

Fig. 5.19 Estimate of the hazard function $\widehat{\lambda}^{(0)}(\cdot, h)$ (solid line), the confidence intervals (dashed line) for ALL data.

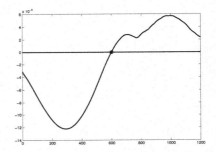

Fig. 5.20 Estimate of $\widehat{\lambda}^{(2)}(\cdot, h)$ for ALL data.

Table 5.5   ALL data.

number of patients	n	=	38
max. follow-up	S	=	2081 days
number of deaths	n_d	=	24
percentage of deaths	p_d	=	63.2%
bandwidths	$\hat{h}_{opt,0,4}$	=	735.86
	$\hat{h}_{opt,2,4}$	=	656.31
most rapid change	$\hat{\theta}_1$	=	599.52

5.6   Use of MATLAB toolbox

The toolbox can be downloaded from the web page
http://www.math.muni.cz/english/science-and-research/
developed-software/232-matlab-toolbox.html.

5.6.1 *Running the program*

The program can be launched by command `kshazard`. If we launch it without parameters it is possible only input data (button (1)) or terminate the program (button (2)) Fig. 5.21).

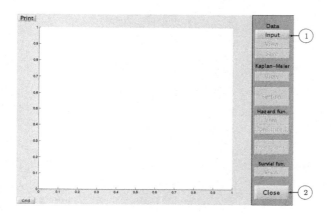

Fig. 5.21 Start of the program.

There is no simulation in the data input window (Fig. 5.22) as the creating of simulated censored data for given hazard function is rather complicated (see Sec. 5.7). We can select the source of the data (the workspace by button (3) or the file by button (4)). Then we choose the names of used variables containing the lifetimes and censoring indicator (buttons (5) and (6)). Finally confirm the values or cancel the subroutine (buttons (7) and (8)).

5.6.2 *Main figure*

Data input causes the same situation as if the toolbox is called with parameters (*e.g.*, `kshazard(X,d)`). A window with data are displayed, uncensored data is represented by crosses, censored by circles (see Fig. 5.23, button (9)). Button (10) will open the window for data saving.

It is also possible to display Kaplan–Meier estimate of the survival function by button (11) (Fig. 5.24). Button (12) opens the window, where the parameters for kernel estimate of hazard function can be set (see Fig. 5.25).

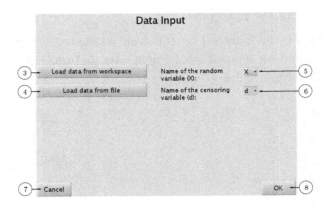

Fig. 5.22   Data input.

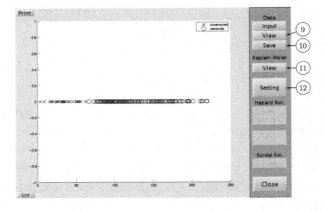

Fig. 5.23   Data view.

5.6.3   *Setting the parameters*

In this subroutine you can choose a predefined kernel (button ⑮), an optimal kernel (button ⑯) and draw the kernel (button ⑰). The optimal kernel can be obtained only for $\nu = 0$. Only three methods for bandwidth selection (button ⑱) are available (see Sec. 5.3).

In boxes ⑳ in the setting window the points for drawing the estimates can be set. Finally we can confirm the selected values by button ㉑ or cancel the subroutine by button ㉒.

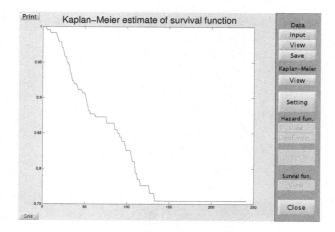

Fig. 5.24   Kaplan–Meier estimate of survival function.

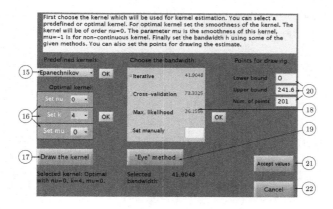

Fig. 5.25   Setting the parameters.

5.6.4   *Eye-control method*

Procedure with the *"Eye-control" method* for bandwidth selection can be invoked by button (19) (see Fig. 5.26). In boxes (23) and (24) we can set the bandwidth and the step for its increasing or decreasing. Pressing the middle button in (25) displays the estimate for the actual bandwidth. The two arrows at the left and the right side cause increasing or decreasing the bandwidth by a step and redrawing the figure. By buttons (26) we can run and stop the gradual increasing the bandwidth and drawing the corre-

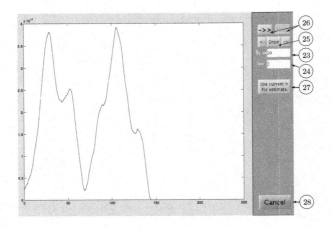

Fig. 5.26   "Eye-control" method.

sponding estimate. Finally it is possible to accept actual bandwidth (button (27)) or cancel the procedure (button (28)) .

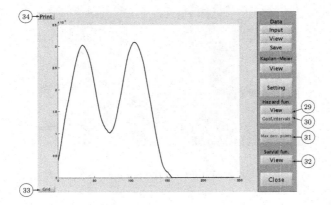

Fig. 5.27   Estimate of the hazard function.

5.6.5 *The final estimation*

After setting the parameters we obtain a window from Fig. 5.27 with the kernel estimate of hazard function (also button (29)). In addition, we have other options: to display the confidence intervals for the kernel estimate of the hazard function (button ③⓪, Fig. 5.28), to show the points of the most rapid change (decreasing) of the estimate of the hazard function (button ③①, Fig. 5.29) and to present the kernel estimate of the survival function (button ③②, Fig. 5.30).

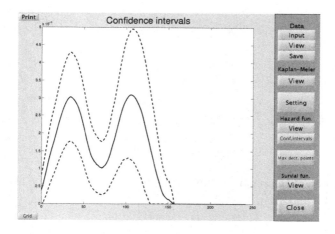

Fig. 5.28   Confidence intervals.

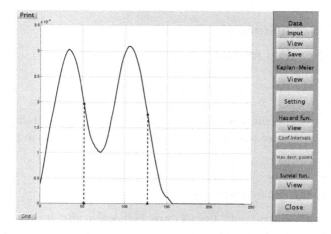

Fig. 5.29   Points of the most rapid decreasing.

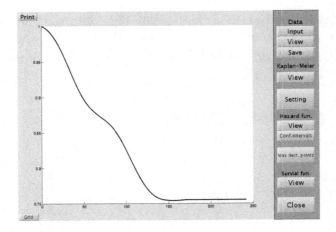

Fig. 5.30   Kernel estimate of the survival function.

The grid and the separate window with the estimate can be also invoked (buttons ③③ and ③④).

5.7   Complements

The following section will focus on simulation of survival data.

Simulation of lifetimes

For the given hazard function λ we have (see (5.2))

$$F(x) = 1 - e^{-\int_0^x \lambda(t)dt}$$

We can see that the lifetimes $T_1,\ldots,T_n$ can be evaluated numerically by re-sampling random variables $U_1,\ldots,U_n$ uniformly distributed on the interval $[0,1]$.

Simulation of censoring times

Assuming the real situation: Let us have a clinical study dealing with some disease. The study starts at time t_0 (we can suppose $t_0 = 0$). Patients enter the study randomly in the interval $[t_0, t_1]$, the beginning of the treatment is given by a random B with the distribution function H. Patient's entry into the study is stopped at time t_1, but the study may continue to some time $t_2 \geq t_1$ when it is finished.

The censoring time is $C = t_2 - B$. For the survival function $\bar{G}$ we have

$$\bar{G}(x) = H(t_2 - x),$$

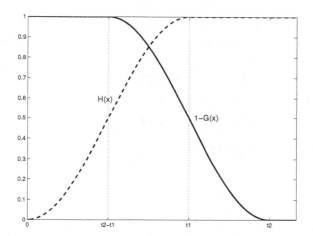

Fig. 5.31 Distribution function for the arrival of patients (H) and the survival function of censoring times ($\bar{G} = 1 - G$)

Let us recall (see (5.10))

$$h_{opt,0,k}^{2k+1} = \frac{\mathcal{L}}{2nkD_k} \gamma_{0,k}^{2k+1},$$

where

$$\mathcal{L} = \int_0^S \frac{\lambda(x)}{\bar{F}(x)\bar{G}(x)} dx.$$

With respect to the choice of values S, the natural choice is $S = t_2$, but this yields the problem with counting $\mathcal{L}$ as $\bar{G}(t_2) = 0$ and the assumptions of Theorem 5.1 are not satisfied. This problem can be solved by choosing $S < t_2$ close to t_2, otherwise, we can use the following procedure:
For the given $\bar{G}$ let us take such λ that

$$\bar{F}(t_2) > 0, \quad \frac{\lambda(x)}{\bar{G}(x)} = O(1), \text{ for } x \to t_2.$$

As a result we have $\lambda(t_2) = 0$ and for $\lambda \in C^{k_0}[0,S]$ also $\lambda'(t_2) = 0$ as λ is a nonnegative function.

In the next example let the beginnings of treatment B be uniformly distributed on $[0, t_1]$. Due to this fact the distribution function C is uniformly distributed on $[t_2 - t_1, t_2]$.

Example 5.1. We use a unimodal hazard function λ on $[0, t_2]$: $\lambda(x) = x(2 - x)^2$, *i.e.*, $F(x) = 1 - e^{\frac{x^2}{12}(3x^2 - 16x + 24)}$. The shape of the hazard function was chosen the same as for the real data – the probability of death initially increases and then decreases. Let K be the Epanechnikov kernel and $n = 100$.

Case A: $t_1 = 1$, $t_2 = 2$, $h_{opt,0,2} = 0.4437$

Case B: $t_1 = 1.5$, $t_2 = 2$, $h_{opt,0,2} = 0.4721$

Case C: $t_1 = 2$, $t_2 = 2$, $h_{opt,0,2} = 0.4993$

The estimates of λ in the particular cases are shown in Fig. 5.32. Figure 5.33

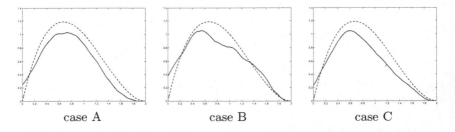

| case A | case B | case C |

Fig. 5.32   Estimates of λ for simulated data: λ – dashed line, estimate – solid line

presents the simulation results of bandwidth estimate for 200 repetitions.

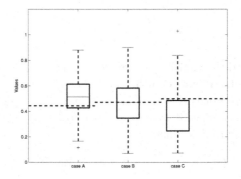

Fig. 5.33   Dashed lines: optimal bandwidths

Chapter 6

Kernel estimation of a regression function

The aim of regression analysis is to produce a reasonable analysis of an unknown regression function m. By reducing the observational errors it allows the interpretation to concentrate on important details of the mean dependence of Y on X.

The task of approximating the regression function can be done essentially in two ways. The quite often used parametric approach is to assume that the mean curve m has some prespecified functional form, for example, a line with an unknown slope and intercept. The literature on the parametric approach is quite extensive, see, *e.g.*, Scheffé (1959); Seber (1977); Searle (1987). But the pure parametric approach often does not meet the need for flexibility in data analysis.

As parametric modeling encounters fundamental difficulties, an attractive alternative are nonparametric curve estimations. In contrast to parametric modeling, the assumptions on the function to be estimated are much weaker, namely only smoothness and differentiability requirements.

However, nonparametric and parametric regression should not be viewed as mutually exclusive competitors. In many cases a nonparametric regression estimate will suggest a simple parametric model, while in other cases it will be clear that the underlying regression function is sufficiently complicated that no reasonable parametric model would be adequate.

As concerns nonparametric methods let us remind that in 1857 the Saxonian economist Engel found the famous "Engelsches Gesetz" (Engel (1857)) by constructing a curve which we would nowadays call a regressogram. There now exist many methods for obtaining a nonparametric estimate of m. Some of the most popular are those based on kernel functions, spline functions and wavelets. Each of these approaches has its own particular strengths and weakness, although kernel regression estimators have the advantage of mathematical and intuitive simplicity. In the context of the kernel regression traditional the Nadaraya – Watson estimator (Nadaraya (1964); Watson (1964)), the Priestley – Chao estimator (Priestley and Chao (1972)), local polynomial kernel estimators (Stone (1977); Cleveland (1979)), the Gasser – Müller estimator (Gasser and Müller (1979)) were developed.

The idea of smoothing was nicely expressed in Eubank (1988), p. 7:

"If m is believed to be smooth, then the observations at X_i near x should contain information about the value of m at x. Thus it should be possible to use something like a local average of data near x to construct an estimator of $m(x)$."

6.1   Basic definition

Consider a standard regression model of the form

$$Y_i = m(x_i) + \varepsilon_i, \qquad i = 1, \dots, n, \tag{6.1}$$

where m is an unknown regression function, $Y_1, \dots, Y_n$ are observable data variables with respect to the design points $x_1, \dots, x_n$. The residuals $\varepsilon_1, \dots, \varepsilon_n$ are independent identically distributed random variables for which

$$\mathrm{E}(\varepsilon_i) = 0, \qquad \mathrm{var}(\varepsilon_i) = \sigma^2 > 0, \qquad i = 1, \dots, n.$$

Remark 6.1. Considering the MATLAB toolbox and its practical usage we will suppose the *fixed equally spaced design*, i.e., design variables are not random and $x_i = i/n$, $i = 1, \dots, n$. In the case of *random design*, where the design points $X_1, \dots, X_n$ are random variables with the same density f, all considerations are similar as for the fixed design. More detailed description of the random design can be found, e.g., in Wand and Jones (1995).

Example 6.1. Figure 6.1 shows a simulated example for which the sample is generated according to the relation

$$Y_i = -6 \frac{\sin(11\, x_i + 5)}{\cot(x_i - 7)} + \varepsilon_i, \qquad i = 1, \dots, 100,$$

where ε_i are drawn from the normal distribution $N(0, 0.2^2)$.

We apply a parametric approach and assume that the underlying curve is linear. But Fig. 6.2 shows that this assumption is not appropriate.

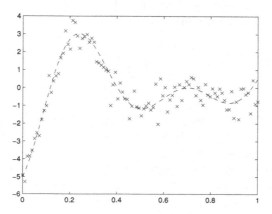

Fig. 6.1 Simulated data with the true regression function (dashed line) from Example 6.1.

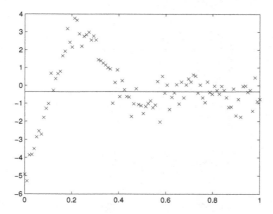

Fig. 6.2 Regression line for simulated data from Example 6.1.

The aim of kernel smoothing is to find a suitable approximation $\widehat{m}$ of the unknown function m. The natural estimate of m, from the given observations (x_i, Y_i), $i = 1, \ldots, n$, is a *regressogram* (see Tukey (1961)). The regressogram is based on the same ideas as the histogram is for density estimation. It consists in dividing the set of values of design points $x_1, \ldots, x_n$ into subintervals B_j, $j = 1, \ldots, r$ and then averaging the values of $Y_1, \ldots, Y_n$ inside each subinterval B_j. The consequent regression estimate is therefore a step function defined for any $x \in B_j$ by

$$\widehat{m}_R(x) = \frac{\sum\limits_{i=1}^{n} Y_i I_{B_j}(x_i)}{\sum\limits_{i=1}^{n} I_{B_j}(x_i)}.$$

This is a typical nonparametric estimate in the sense that no parametric form needs to be assumed. However, this estimate has the disadvantage that both the number and the position of the intervals B_j to be chosen. Figure 6.3 shows the regressogram of the regression function from Example 6.1 obtained by dividing the interval $[0, 1]$ into 12 equal subintervals. The dashed line represents the true regression function, the full line represents the regressogram.

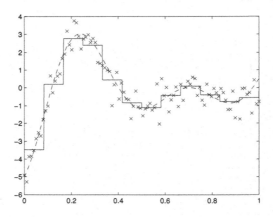

Fig. 6.3   Regressogram for simulated data from Example 6.1.

The first natural extension of the regressogram is a *moving averages* estimate. The idea is that a local average of Y_i values is still used, but the estimation at the point x depends on values from a centered neighborhood of x. More precisely, it is defined by

$$\widehat{m}_W(x) = \frac{\sum\limits_{i=1}^{n} Y_i I_{[x-h,x+h]}(x_i)}{\sum\limits_{i=1}^{n} I_{[x-h,x+h]}(x_i)}.$$

The positive parameter h controls the size of the neighborhood around x and thus it has a great influence on the behavior of the estimate.

Finally, this estimate can be generalized by using a weighted average by means of some kernel function K, $K \in S_{0,k}$, k is even. It is known as the *Nadaraya – Watson estimator* (Nadaraya (1964) and Watson (1964)) and the estimate at the point x, $h < x < 1 - h$, is defined by

$$\widehat{m}_{NW}(x,h) = \frac{\sum\limits_{i=1}^{n} K_h(x_i - x)Y_i}{\sum\limits_{i=1}^{n} K_h(x_i - x)}. \tag{6.2}$$

Figure 6.4 demonstrates the construction of Nadaraya – Watson weights for the given point x (see also Seifert and Gasser (1996)).

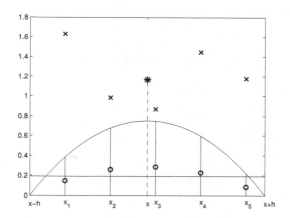

Fig. 6.4   Illustration of the Nadaraya – Watson estimator.

The crosses represent five observations (x_1, Y_1), ...,(x_5, Y_5), the parabola is the scaled Epanechnikov kernel K_h and the vertical lines are its values at the design points x_i. These values are divided by the constant $\sum\limits_{i=1}^{5} K_h(x_i - x)$ which is represented by the horizontal line. The circles stand for the resulting weights

$$W_i(x,h) = \frac{K_h(x_i - x)}{\sum\limits_{i=1}^{5} K_h(x_i - x)}, \quad i = 1, \ldots, 5.$$

The final estimate $\widehat{m}(x)$ is represented by the star.

In order to complete the overview of commonly used nonparametric methods for estimating $m(x)$ we mention these estimators:

- *local – linear estimator* (Stone (1977); Cleveland (1979))

$$\widehat{m}_{LL}(x,h) = \frac{1}{n}\sum_{i=1}^{n}\frac{\{\hat{s}_2(x,h) - \hat{s}_1(x,h)(x_i - x)\}K_h(x_i - x)Y_i}{\hat{s}_2(x,h)\hat{s}_0(x,h) - \hat{s}_1(x,h)^2}, \qquad (6.3)$$

where

$$\hat{s}_r(x,h) = \frac{1}{n}\sum_{i=1}^{n}(x_i - x)^r K_h(x_i - x), \qquad r = 0,1,2,$$

- *Priestley – Chao estimator* (Priestley and Chao (1972))

$$\widehat{m}_{PCH}(x,h) = \frac{1}{n}\sum_{i=1}^{n}K_h(x_i - x)Y_i, \qquad (6.4)$$

- *Gasser – Müller estimator* (Gasser and Müller (1979))

$$\widehat{m}_{GM}(x,h) = \sum_{i=1}^{n}Y_i\int_{s_{i-1}}^{s_i}K_h(t - x)dt, \qquad (6.5)$$

where

$$s_i = \frac{x_i + x_{i+1}}{2} = \frac{2i + 1}{2n}, \quad i = 1,\ldots,n-1, \quad s_0 = 0, \quad s_n = 1.$$

One can see from these formulas that kernel estimators can be generally expressed as

$$\widehat{m}(x,h) = \sum_{i=1}^{n}W_i^{(j)}(x,h)Y_i, \qquad (6.6)$$

where the weights $W_i^{(j)}(x,h)$, $j \in \{NW, LL, PCH, GM\}$ correspond to the weights of estimators $\widehat{m}_{NW}$, $\widehat{m}_{LL}$, $\widehat{m}_{PCH}$ and $\widehat{m}_{GM}$ defined above. For the sake of simplicity we will write only $W_i(x,h)$ further in the text.

6.2 Statistical properties of the estimate

The quality of a kernel regression estimator can be locally described by the Mean Square Error (MSE)

$$\text{MSE}\{\widehat{m}(x,h)\} = E\{\widehat{m}(x,h) - m(x)\}^2.$$

It could be also expressed as a decomposition into a variance and a squared bias

$$\text{MSE}\{\widehat{m}(x,h)\} = \text{var}\{\widehat{m}(x,h)\} + \text{bias}^2\{\widehat{m}(x,h)\}.$$

Remark 6.2. We mention some asymptotic properties of the Mean Square Error at an "inner" point x of the interval $[0,1]$. Since the estimators $\widehat{m}_{NW}, \widehat{m}_{LL}, \widehat{m}_{PCH}, \widehat{m}_{GM}$ are asymptotically equivalent in this case (see, *e.g.*, Lejeune (1985); Müller (1987); Wand and Jones (1995)), the lower indices will be left out and we will write only $\widehat{m}$ for the kernel regression estimator.

We make the following assumptions for our analysis:

- $K \in S_{0,k}$ and k is even,
- $m \in C^{k_0}[0,1]$, $k_0 > k$,
- $h = h(n)$ is a sequence satisfying $\lim\limits_{n\to\infty} h = 0$ and $\lim\limits_{n\to\infty} nh = \infty$,
- the point x is an inner point of the interval $[0,1]$, *i.e.*, there exists an index n_0 such that

$$h < x < 1 - h, \ \forall n \geq n_0.$$

Under the above assumptions, the formulas for bias$\{\widehat{m}(x,h)\}$ and var$\{\widehat{m}(x,h)\}$ can be expressed. Results for the local linear (and generally for the local polynomial) estimator with $K \in S_{0,2}$ can be found, *e.g.*, in Wand and Jones (1995). Koláček (2005) obtained results for the Nadaraya – Watson estimator with $k \geq 2$. These formulas take the forms

$$\text{bias}\{\widehat{m}(x,h)\} = \frac{\beta_k}{k!} m^{(k)}(x) h^k + o(h^k) + O(n^{-1}), \qquad (6.7)$$

$$\text{var}\{\widehat{m}(x,h)\} = \frac{\sigma^2}{nh} V(K) + o(n^{-1}h^{-1}). \qquad (6.8)$$

The Mean Square Error of $\widehat{m}$ for the random design and for the boundary points is discussed in many papers, *e.g.*, in Mack and Müller (1988); Fan (1992); Chu and Marron (1991); Wand and Jones (1995) for all types of mentioned estimators.

As concerns a global criterion the Mean Integrated Square Error (MISE) is considered

$$\text{MISE}\{\widehat{m}(\cdot,h)\} = E \int\limits_0^1 \{\widehat{m}(x,h) - m(x)\}^2 dx = \int\limits_0^1 \text{MSE}\{\widehat{m}(x,h)\} dx.$$

Theorem 6.1. *Let assumptions given above be satisfied. Then*

$$\text{MISE}\{\widehat{m}(\cdot,h)\} = \frac{\sigma^2}{nh} V(K) + \left(\frac{\beta_k}{k!}\right)^2 A_k h^{2k} + o\left\{h^{2k} + (nh)^{-1}\right\}, \quad (6.9)$$

where $A_k = \int \left(m^{(k)}(x)\right)^2 dx.$

Proof. The formulas (6.7) and (6.8) yield the result. $\qquad\square$

Since MISE is not mathematically tractable we employ the Asymptotic Mean Integrated Square Error (AMISE) which can be written as a sum of the Asymptotic Integrated Variance and Asymptotic Integrated Square Bias

$$\text{AMISE}\{\widehat{m}(\cdot,h)\} = \underbrace{\frac{V(K)\sigma^2}{nh}}_{\text{AIV}} + \underbrace{\left(\frac{\beta_k}{k!}\right)^2 A_k h^{2k}}_{\text{AISB}}. \qquad (6.10)$$

As in the density case, AMISE can be expressed in the following way:

$$\text{AMISE}\{\widehat{m}(\cdot,h)\} = T(K)\left(\frac{\gamma_{0k}\sigma^2}{nh} + \frac{h^{2k}A_k}{\gamma_{0k}^{2k}(k!)^2}\right), \qquad (6.11)$$

where the functional $T(K)$ is defined by (1.3) and γ_{0k} is defined by (1.4). The optimal bandwidth considered here is $h_{opt,0,k}$, the minimizer of (6.10), *i.e.*,

$$h_{opt,0,k} = \arg\min_{h\in H_n} \text{AMISE}\{\widehat{m}(\cdot,h)\},$$

where $H_n = [an^{-1/(2k+1)}, bn^{-1/(2k+1)}]$ for some $0 < a < b < \infty$. The calculation gives

$$h_{opt,0,k} = \left(\frac{\sigma^2 V(K)(k!)^2}{2kn\beta_k^2 A_k}\right)^{\frac{1}{2k+1}}, \qquad (6.12)$$

i.e., $h_{opt,0,k} = O(n^{-\frac{1}{2k+1}})$.

To avoid the numerical integration in practice for $\text{MISE}\{\widehat{m}(\cdot,h)\}$ it seems to be more appropriate to analyze the Average Mean Square Error

$$\text{AMSE}\{\widehat{m}(\cdot,h)\} = \frac{1}{n}E\sum_{i=1}^{n}\{m(x_i) - \widehat{m}(x_i,h)\}^2. \qquad (6.13)$$

Remark 6.3. *(The confidence interval)*
The confidence interval is very useful in many applications. It takes the form

$$\left[\widehat{m}(x,h) - u_{1-\frac{\alpha}{2}}\sqrt{\frac{V(K)\widehat{\sigma}^2(x)}{nh}}, \widehat{m}(x,h) + u_{1-\frac{\alpha}{2}}\sqrt{\frac{V(K)\widehat{\sigma}^2(x)}{nh}}\right], \quad (6.14)$$

where $u_{1-\frac{\alpha}{2}}$ is the $(1-\alpha/2)$-quantile of the standard normal distribution Φ and the variance estimate $\widehat{\sigma}^2(x)$ at x is given by

$$\widehat{\sigma}^2(x) = \sum_{i=1}^{n} W_i(x,h)\{Y_i - \widehat{m}(x,h)\}^2.$$

The detailed construction of confidence intervals for $\widehat{m}$ can be found, *e.g.*, in Härdle (1990).

Remark 6.4. *(Choosing the shape of the kernel)*
We assume $K \in S^0_{0,k}$ and under additional assumption that k is even, $k > 0$, we can proceed in the same way as in Sec. 2.3.

6.3 Choosing the bandwidth

In nonparametric regression estimation, like in density estimation, a critical and inevitable step is to choose the smoothing parameter (bandwidth) to control the smoothness of the curve estimate. The smoothing parameter considerably affects the features of the estimated curve. It is shown on the simulated data from Example 6.1. We note that the small values of h (see Fig. 6.5(a)) produce an undersmoothed estimate while the values of h that are too large (see Fig. 6.5(c)) lead to an oversmoothed estimate. Apparently, Fig. 6.5(b) presents the optimal kernel estimate.

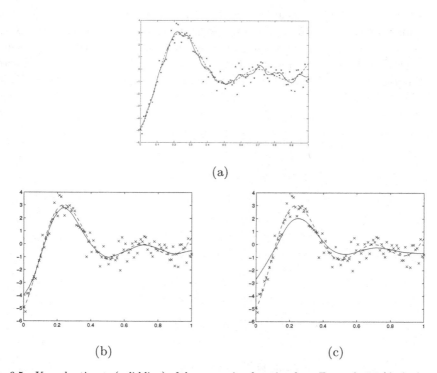

(a)

(b)   (c)

Fig. 6.5   Kernel estimate (solid line) of the regression function from Example 6.1 (dashed line) for 3 different bandwidths: (a) $h = 0.03$, (b) $h = 0.08$, (c) $h = 0.14$.

Although in practice one can try several bandwidths and choose a bandwidth subjectively, automatic (data-driven) selection procedures could be useful for many situations; see Silverman (1985) for more examples.

Most of these procedures are based on estimating of $\text{AMSE}\{\widehat{m}(\cdot, h)\}$ defined in (6.13). They are asymptotically equivalent and asymptotically unbiased (see Härdle (1990); Chiu (1990, 1991)). However, in simulation studies (Koláček (2005)), it is often observed that most selectors are biased toward undersmoothing and yield smaller bandwidths more frequently than predicted by asymptotic results.

A naive estimate of $\text{AMSE}\{\widehat{m}(\cdot, h)\}$ is the Residual Sum of Squares (RSS)

$$\text{RSS}_n(h) = \frac{1}{n} \sum_{i=1}^{n} \{Y_i - \widehat{m}(x_i, h)\}^2, \qquad (6.15)$$

where the unknown mean $m(x_i)$ is replaced by the observation Y_i at x_i. Unfortunately, $\text{RSS}(h)$ is a biased estimate of $\text{AMSE}\{\widehat{m}(\cdot, h)\}$. Figure 6.6 shows $\text{RSS}(h)$ as an increasing function in h, *i.e.*, the optimal bandwidth would be the smallest bandwidth. However, the $\text{RSS}(h)$ is the initial error function for all bandwidth selectors described in the following paragraphs.

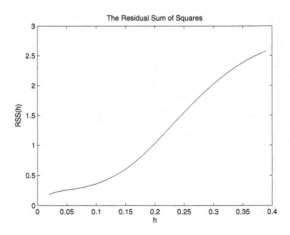

Fig. 6.6   Residual Sum of Squares for simulated data from Example 6.1.

6.3.1 *Mallows' method*

First, it is necessary to explain why the Residual Sum of Squares is a biased estimate of AMSE. It is described in Theorem 6.2 precisely. The RSS function can be expressed as

$$
\text{RSS}_n(h) = \frac{1}{n} \sum_{i=1}^{n} \{\widehat{m}(x_i, h) - m(x_i)\}^2
$$
$$
+ \frac{1}{n} \sum_{i=1}^{n} \varepsilon_i^2 - \frac{2}{n} \sum_{i=1}^{n} \varepsilon_i \left[\sum_{j=1}^{n} W_j(x_i, h) Y_j - m(x_i) \right]. \tag{6.16}
$$

Set $C_{1n}(h)$ the last term in (6.16), *i.e.*,

$$
C_{1n}(h) = \frac{2}{n} \sum_{i=1}^{n} \varepsilon_i \left[\sum_{j=1}^{n} W_j(x_i, h) Y_j - m(x_i) \right].
$$

The following theorem emphasizes the fact that the expectation of $C_{1n}(h)$ is nonzero.

Theorem 6.2. *The expected value of* $\text{RSS}_n(h)$ *equals*

$$
E(\text{RSS}_n(h)) = \text{AMSE}\{\widehat{m}(\cdot, h)\} + \sigma^2 - \frac{2\sigma^2}{n} \sum_{i=1}^{n} W_i(x_i, h).
$$

Thus RSS_n *is a biased estimate of AMSE.*

Proof. It is an easy exercise to prove the result. □

Rice (1984) considered the error function

$$
\widehat{R}_n(h) = \text{RSS}_n(h) - \hat{\sigma}^2 + \frac{2\hat{\sigma}^2}{n} \sum_{i=1}^{n} W_i(x_i, h), \tag{6.17}
$$

where $\hat{\sigma}^2$ is an estimate of σ^2

$$
\hat{\sigma}^2 = \frac{1}{2n - 2} \sum_{i=2}^{n} (Y_i - Y_{i-1})^2. \tag{6.18}
$$

The estimate $\hat{h}_{\text{M}}$ of the optimal bandwidth is defined as

$$
\hat{h}_{\text{M}} = \underset{h \in H_n}{\arg \min} \widehat{R}_n(h).
$$

A similar approach was proposed in Mallows (1973). Thus comes the name of this method.

6.3.2   *Cross-validation method*

One of the most widespread procedures for bandwidth selection is the cross-validation method, also known as "leave-one-out" method. The literature on this criterion is quite extensive, *e.g.*, Stone (1974); Craven and Wahba (1979); Härdle (1990); Droge (1996). The idea is similar as in density estimations.

The method is based on modified regression smoothers (6.6) in which one, say the j-th, observation is left out:

$$\widehat{m}_{-j}(x_j, h) = \sum_{\substack{i=1 \\ i \neq j}}^{n} W_i(x_j, h) Y_i.$$

With using these modified smoothers, the function $\mathrm{RSS}_n(h)$ takes the form

$$\mathrm{CV}(h) = \frac{1}{n} \sum_{i=1}^{n} \{\widehat{m}_{-i}(x_i) - Y_i\}^2. \tag{6.19}$$

The function $\mathrm{CV}(h)$ is commonly called a "cross-validation" function. Let $\hat{h}_{\mathrm{CV}}$ stand for minimization of $\mathrm{CV}(h)$, *i.e.*,

$$\hat{h}_{\mathrm{CV}} = \operatorname*{arg\,min}_{h \in H_n} \mathrm{CV}(h).$$

The reason why cross-validation works is simple. The CV function can be expressed in the form

$$\mathrm{CV}(h) = \frac{1}{n} \sum_{i=1}^{n} \{\widehat{m}_{-i}(x_i, h) - m(x_i)\}^2$$

$$+ \frac{1}{n} \sum_{i=1}^{n} \varepsilon_i^2 - \frac{2}{n} \sum_{i=1}^{n} \varepsilon_i \left[\sum_{\substack{j=1 \\ j \neq i}}^{n} W_j(x_i, h) Y_j - m(x_i) \right]. \tag{6.20}$$

By using (6.20), it is easy to prove the statement of the following theorem.

Theorem 6.3. *The expected value of* $\mathrm{CV}(h)$ *equals*
$$E(\mathrm{CV}(h)) = \mathrm{AMSE}\{\widehat{m}(\cdot, h)\} + \sigma^2.$$

Remark 6.5. Set the last term in (6.20) as $C_{2n}(h)$. Let us note that the fact that $C_{2n}(h)$ has expectation zero does not guarantee that $\hat{h}_{\mathrm{CV}}$ minimizes AMSE (or any other of the equivalent error measures). For the cross-validation procedure it is necessary that $C_{2n}(h)$ converges uniformly over h to zero. However, in practice, this term does not often satisfy the condition and affects the bias of $\mathrm{CV}(h)$. Mostly, its minimum $\hat{h}_{\mathrm{CV}}$ takes less values than the optimal bandwidth.

6.3.3 *Penalizing functions*

The third proposal, based on modifying $\text{RSS}_n(h)$ in a suitable way, aims at an asymptotic cancellation of the bias of $\text{RSS}_n(h)$. This idea was discussed, *e.g.*, in Härdle *et al.* (1988); Härdle (1990). They introduced the "penalizing" function $\Xi(u)$ for this purpose.

Definition 6.1. Any function $\Xi(u)$, for which the first-order Taylor expansion at zero takes the form

$$\Xi(u) = 1 + 2u + O(u^2), \tag{6.21}$$

is called a *penalizing* function.

This form of the penalizing function works out well as it will be evident from the following considerations. The i-th term of $\text{RSS}_n(h)$ is adjusted by $\Xi(n^{-1}W_i(x_i, h))$, that is, modified to

$$G(h) = \frac{1}{n}\sum_{i=1}^{n}\{\widehat{m}(x_i, h) - Y_i\}^2 \Xi(n^{-1}W_i(x_i, h)). \tag{6.22}$$

The reason for this adjustment is that the correction function $\Xi(n^{-1}W_i(x_i, h))$ penalizes small values of h. Using the Taylor expansion (6.21) for Ξ in (6.22) and disregarding the higher order terms leads to the expression

$$\begin{aligned}
G(h) = &\frac{1}{n}\sum_{i=1}^{n}\{\widehat{m}(x_i, h) - m(x_i)\}^2 \\
&+ \frac{1}{n}\sum_{i=1}^{n}\varepsilon_i^2 - C_{1n}(h) + \frac{2}{n}\sum_{i=1}^{n}\varepsilon_i^2\frac{1}{n}W_i(x_i, h),
\end{aligned} \tag{6.23}$$

where $C_{1n}(h)$ is defined in §6.3.1. Note that the second term is independent of h and that the expectation of $C_{1n}(h)$ is equal to the expectation of the last term (see Theorem 6.2). So the last two terms vanish asymptotically as it can be seen in the following theorem.

Theorem 6.4. *The expected value of* $G(h)$ *equals*

$$E(G(h)) = \text{AMSE}\{\widehat{m}(\cdot, h)\} + \sigma^2.$$

Proof. The result follows from the expression (6.23) and from Theorem 6.2. $\qquad\square$

The estimate $\hat{h}_{\text{PEN}}$ of the optimal bandwidth is defined as

$$\hat{h}_{\text{PEN}} = \arg\min_{h \in H_n} G(h).$$

Example 6.2. Some examples of penalizing functions are listed below. These functions are used in the MATLAB toolbox.

(1) *Generalized cross-validation* (Craven and Wahba (1979); Li (1985))

$$\Xi_{GCV}(u) = \frac{1}{(1-u)^2}.$$

(2) *Akaike's Information Criterion* (Akaike (1970))

$$\Xi_{AIC}(u) = e^{2u}.$$

(3) *Finite Prediction Error* (Akaike (1974))

$$\Xi_{FPE}(u) = \frac{1+u}{1-u}.$$

(4) *Shibata's model selector* (Shibata (1981))

$$\Xi_S(u) = 1 + 2u.$$

(5) *Rice's bandwidth selector* (Rice (1984))

$$\Xi_R(u) = \frac{1}{1-2u}.$$

(6) *ET bandwidth selector* (Koláček (2002))

$$\Xi_{ET}(u) = e^{\frac{2}{\pi}\tan \pi u}.$$

All these penalizing functions are illustrated in Fig. 6.7. The comparison of all mentioned penalizing functions in simulation study and application to real data can be found in Koláček (2005).

Note. The cross-validation method can be considered as penalizing the prediction error $\text{RSS}_n(h)$, since

$$\frac{\text{CV}(h)}{\text{RSS}_n(h)} = 1 + 2n^{-1}W_i(x_i, h) + O(n^{-2}h^{-2}).$$

For more details see Härdle *et al.* (1988).

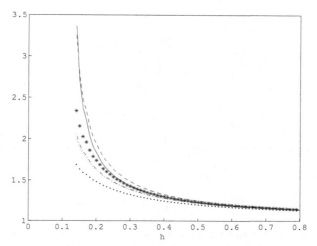

Fig. 6.7 Penalizing functions from Example 6.2: - - Rice, — ET, ** Generalized, — · — FPE, .. Akaike, •• Shibata.

6.3.4 *Methods based on Fourier transformation*

In the simulation study of Chiu (1990), it was observed that Mallows' criterion (see §6.3.1) gives smaller bandwidths more frequently than predicted by the asymptotic theorems. Chiu (1990) provided an explanation for the cause and suggested a procedure to overcome the difficulty. By applying the procedure, we introduce a method for bandwidth selection which gives much more stable bandwidth estimates. As an outgrowth of ideas used in this approach we obtain a type of plug-in method, *i.e.*, an estimate of the constant A_k defined in Theorem 6.1.

Notation 6.1. We assume a "cyclic design", that is, suppose m is a smooth periodic function and the estimate is obtained by applying the kernel on the extended series $\widetilde{Y}_i$, $i = -n + 1, -n + 2, \ldots, 2n$, where generally $\widetilde{Y}_{j+ln} = Y_j$ for $j = 1, \ldots, n$ and $l \in \mathbb{Z}$. Similarly $x_i = i/n$, $i = -n + 1, -n + 2, \ldots, 2n$.

Remark 6.6. The assumption of the cyclic model leads to the fact that the weights of Nadaraya – Watson and local linear estimator are identical at the design points that is

$$W_i^{(LL)}(x_j, h) = W_i^{(NW)}(x_j, h),$$

for $i \in \{-n + 1, -n + 2, \ldots, 2n\}$, $j \in \{1, 2, \ldots, n\}$. In the following, we will write only $W_i(x_j, h)$ without the upper index and the kernel estimators will be expressed as

$$\widehat{m}(x,h) = \sum_{i=-n+1}^{2n} W_i(x,h)\widetilde{Y}_i.$$

More precisely, let $K \in S_{0,k}$, $h \in (0,1)$, $j \in \{1,\ldots,n\}$. Then the sum $\sum_{i=-n+1}^{2n} K_h(x_i - x_j) = \sum_{i=-n+1}^{n-1} K_h(x_i)$ is independent on j. Set $B_n = \sum_{i=-n+1}^{n-1} K_h(x_i)$. We can simply evaluate weight functions at the design points x_j, $j = 1,\ldots,n$

$$W_i(x_j,h) = \frac{1}{B_n}K_h(x_i - x_j). \tag{6.24}$$

Let us remark that Chiu's procedure was proposed for Priestley – Chao estimator and for a special class of symmetric probability density functions from $S_{0,2}$ as kernels. We follow the Nadaraya – Watson and local – linear estimators especially and extend the procedure to these estimators and to kernels from the class $S_{0,k}$, k even.

In the further text, we need some notations and definitions from the discrete Fourier theory. Let us denote $\mathbf{Y} = (Y_1,\ldots,Y_n)^T$ the vector of observations.

Definition 6.2. The *periodogram* of $\mathbf{Y}$ is defined by $\mathbf{I}_Y = (I_{Y_1},\ldots,I_{Y_n})^T$

$$I_{Y_j} = |Y_j^-|^2/2\pi n, \quad j = 1,\ldots,n,$$

where

$$Y_j^- = \sum_{s=0}^{n-1} Y_{s+1}e^{-\frac{i2\pi s(j-1)}{n}}$$

is the finite Fourier transform of the vector $\mathbf{Y}$. This transformation is denoted by $\mathbf{Y}^- = DFT^-(\mathbf{Y})$.

Let us denote $\mathbf{m} = (m(x_1),\ldots,m(x_n))^T$ and $\boldsymbol{\varepsilon} = (\varepsilon_1,\ldots,\varepsilon_n)^T$. The periodograms and Fourier transforms of the vectors $\boldsymbol{\varepsilon}$ and $\mathbf{m}$ are defined similarly. Under mild conditions, the periodogram ordinates I_{ε_j} on Fourier frequencies $2\pi(j-1)/n$, for $j = 2,\ldots,N = \left[\frac{n}{2}\right]$, are approximately independently and exponentially distributed with means $\sigma^2/(2\pi)$. See Brillinger (2001) for a more precise statement. Here $[x]$ means the greatest integer less or equal to x.

Definition 6.3. Let $\mathbf{u} = (u_1, \ldots, u_n)^T, \mathbf{v} = (v_1, \ldots, v_n)^T \in \mathbb{C}^n$;

$$z_j = \sum_{i=1}^{n} u_{\langle j-i+1 \rangle_n} v_i,$$

where $\langle j - i + 1 \rangle_n$ marks $(j - i + 1) \mod n$. Then $\mathbf{z} = (z_1, \ldots, z_n)^T$ is called *the discrete cyclic convolution* of vectors $\mathbf{u}$ and $\mathbf{v}$; we write $\mathbf{z} = \mathbf{u} \circledast \mathbf{v}$.

Let us denote a vector $\mathbf{w} = (w_1, w_2, \ldots, w_n)^T$, where

$$w_j = W_0(x_{j-1} - 1, h) + W_0(x_{j-1}, h) + W_0(x_{j-1} + 1, h), \ j \in \{1, \ldots, n\}.$$

Then we express $\widehat{m}(x_j, h)$ as a discrete cyclic convolution of vectors $\mathbf{w}$ and $\mathbf{Y}$

$$\widehat{m}(x_j, h) = \sum_{i=1}^{n} w_{\langle j-i+1 \rangle_n} Y_i, \tag{6.25}$$

i.e., $\widehat{\mathbf{m}} = \mathbf{w} \circledast \mathbf{Y}$.

The application of Parseval's formula yields

$$\mathrm{RSS}_n(h) = \frac{4\pi}{n} \sum_{j=2}^{N} I_{Y_j} \left\{ 1 - w_j^- \right\}^2, \tag{6.26}$$

where $w_j^- = \sum_{s=-n+1}^{n-1} W_0(x_s, h) e^{-\frac{i 2 \pi s (j-1)}{n}}$ is the finite Fourier transform of $\mathbf{w}$ (see Chiu (1990) for details). The equivalent expression for Mallows' criterion $\widehat{R}_n(h)$ is derived from (6.17) and (6.26)

$$\widehat{R}_n(h) = \frac{4\pi}{n} \sum_{j=2}^{N} I_{Y_j} \{1 - w_j^-\}^2 - \hat{\sigma}^2 + 2\hat{\sigma}^2 w_1. \tag{6.27}$$

Similarly, we arrive at the formula for $\mathrm{AMSE}\{\widehat{m}(\cdot, h)\}$

$$\mathrm{AMSE}\{\widehat{m}(\cdot, h)\} = \frac{4\pi}{n} \sum_{j=2}^{N} \left\{ I_{m_j} + \frac{\sigma^2}{2\pi} \right\} \{1 - w_j^-\}^2 - \sigma^2 + 2\sigma^2 w_1. \tag{6.28}$$

Formulas (6.27) and (6.28) are very useful for our further considerations.

Remark 6.7. *(The motivation)*
Let $D(h) = \widehat{R}_n(h) - \mathrm{AMSE}\{\widehat{m}(\cdot, h)\}$. From the previous expressions we obtain

$$D(h) = \frac{4\pi}{n} \sum_{j=2}^{N} \left\{ I_{Y_j} - I_{m_j} - \frac{\sigma^2}{2\pi} \right\} \{1 - w_j^-\}^2.$$

The periodogram ordinates I_{m_j} for the vector $\mathbf{m} = (m(x_1), \ldots, m(x_n))^T$ decrease rapidly for smooth $m(x)$. Therefore I_{Y_j} do not contain much information about I_{m_j} at high frequencies (for the rigorous proof see Rice (1984)). This leads to the consideration of the procedure proposed by Chiu (1991). The main idea is to modify RSS to make it less variable. We find the first index J_1 such that $I_{Y_{J_1}} < c\hat{\sigma}^2/2\pi$ for some constant $c > 1$, where $\hat{\sigma}^2$ is an estimate of σ^2. The constant c sets a threshold. In our experience, setting $1 < c < 3$ yields good results.

The modified residual sum of squares is defined by

$$\mathrm{MRSS}_n(h) = \frac{4\pi}{n} \sum_{j=2}^{N} \tilde{I}_{Y_j} \{1 - w_j^-\}^2, \qquad (6.29)$$

where

$$\tilde{I}_{Y_j} = \begin{cases} I_{Y_j}, & 2 \le j < J_1 \\ \hat{\sigma}^2/2\pi, & J_1 \le j \le N. \end{cases}$$

See Fig. 6.8 and Fig. 6.9 for an insight. Thus, the proposed selector takes the form

$$\widetilde{R}_n(h) = \mathrm{MRSS}_n(h) - \hat{\sigma}^2 + 2\hat{\sigma}^2 w_1 \qquad (6.30)$$

and the estimate of the optimal bandwidth is

$$\hat{h}_{\mathrm{MRSS}} = \arg\min_{h \in H_n} \widetilde{R}_n(h).$$

For more details see Chiu (1990, 1991).

As the second method in this paragraph, we introduce a plug-in method which is also based on discrete Fourier transformation and uses some properties of transformed error functions from the previous text. The method was developed in Koláček (2008).

To simplify the discussion below, set $c = 2$ and rewrite (6.30) to the formula in the next lemma.

Lemma 6.1. *Let J_1 be the least index such that $I_{Y_{J_1}} < \hat{\sigma}^2/\pi$. Then*

$$\widetilde{R}_n(h) = \frac{\hat{\sigma}^2}{n} \sum_{j=1}^{n} (w_j^-)^2 + \frac{4\pi}{n} \sum_{j=2}^{J_1-1} \left\{ I_{Y_j} - \frac{\hat{\sigma}^2}{2\pi} \right\} \{1 - w_j^-\}^2.$$

Proof. The rigorous proof can be found in Complements. $\qquad \square$

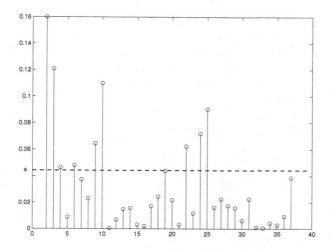

Fig. 6.8   Periodogram ordinates I_{Y_j} as a function of j, $a = 2\frac{\hat{\sigma}^2}{2\pi}$.

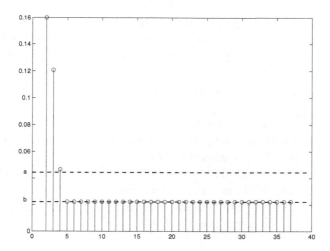

Fig. 6.9   Modified periodogram ordinates $\tilde{I}_{Y_j}$ as a function of j, $a = 2\frac{\hat{\sigma}^2}{2\pi}$, $b = \frac{\hat{\sigma}^2}{2\pi}$.

The main idea of plug-in method is to estimate unknown parameters σ^2 and A_k in the formula (6.12) for the optimal bandwidth $h_{opt,0,k}$, which realizes the minimum of AMISE$\{\widehat{m}(\cdot, h)\}$ (see (6.10))

$$\text{AMISE}\{\widehat{m}(\cdot, h)\} = \frac{\sigma^2 V(K)}{nh} + \left(\frac{\beta_k}{k!}\right)^2 A_k h^{2k}.$$

We can use the formula (6.18) for estimating σ^2. However, in the case of estimation of A_k, the situation is more complicated. According to the previous considerations we can replace the error function AMISE$\{\widehat{m}(\cdot, h)\}$ by the selector $\widetilde{R}_n(h)$ expressed in Lemma 6.1. If we compare these two error functions, we arrive at results described in the next theorems (see Complements for more detailed proofs).

Theorem 6.5. *Let* $\mathbf{w}^-$ *be the discrete Fourier transformation of vector* $\mathbf{w}$. *Then*

$$\sum_{j=1}^{n}(w_j^-)^2 = \frac{V(K)}{h} + o(h^{-1}). \qquad (6.31)$$

The previous theorem implies that the first term of $\widetilde{R}_n(h)$ estimates the variance term of AMISE$\{\widehat{m}(\cdot, h)\}$, *i.e.*,

$$\frac{\hat{\sigma}^2}{n}\sum_{j=1}^{n}(w_j^-)^2 = \frac{\sigma^2 V(K)}{nh} + o\left\{(nh)^{-1}\right\}.$$

Similarly, we can compare the bias terms to obtain an estimator for A_k. Let $\varepsilon > 0$, $h \in (0,1)$, set J_2 the last index from $\{1, \ldots, n\}$ for which

$$J_2 \leq \frac{\sqrt[k+1]{\varepsilon(k+1)!}}{2\pi h}.$$

Let us notice that the parameter ε is an error of Taylor's approximation used in the proof of Theorem 6.6 (see Complements) and the parameter h is some "starting" approximation of $h_{opt,0,k}$. In our experience, setting $\varepsilon = 10^{-3}$ and $h = k/n$ yields good results. Some other considerations and rigorous proofs can be found in Koláček (2008).

Further we request both conditions for indices J_1 and J_2 hold simultaneously, therefore we define the index J

$$J = \min\{J_1, J_2 + 1\}. \qquad (6.32)$$

Theorem 6.6. *Let* J *be the index defined by* (6.32). *Then for all* $j \in \mathbb{N}$, $1 \leq j \leq J - 1$, *it holds*

$$\frac{1}{(2\pi j)^k}(1 - w_{j+1}^-) = (-1)^{\frac{k}{2}+1}\frac{h^k}{k!}\beta_k + \xi + o\left(h^k\right), \qquad (6.33)$$

where ξ *is a constant satisfying* $|\xi| < \varepsilon$.

By using the result of this theorem we can derive the estimator of an unknown parameter A_k. We mention only the result here, see Complements for the main ideas of derivation.

Let J be the index defined by (6.32). Then the estimator of the parameter A_k is of the form

$$\widehat{A}_k = \frac{4\pi}{n} \sum_{j=1}^{J-2} (2\pi j)^{2k} \left\{ I_{Y_{j+1}} - \frac{\hat{\sigma}^2}{2\pi} \right\}. \tag{6.34}$$

By putting this estimator into (6.12) we obtain the plug-in estimator of $h_{opt,0,k}$

$$\hat{h}_{\mathrm{PI}} = \left(\frac{\hat{\sigma}^2 V(K)(k!)^2}{2kn\beta_k^2 \widehat{A}_k} \right)^{\frac{1}{2k+1}}. \tag{6.35}$$

We would like to point out the computational aspect of the plug-in method. It has preferable properties to classical methods because there is no problem of minimization of any error function. Also the sample size which is necessary for computing the estimation is far less than for classical methods. On the other side, a minor disadvantage could be the fact that we need some "starting" approximation of the unknown parameter h. We also remark that both methods in this paragraph are developed for a rather limited case: the cyclic design.

6.4 Estimation of the derivative of the regression function

This paragraph is devoted to the kernel estimate of the ν-th derivative $m^{(\nu)}$ of the regression function m. We start with some additional assumptions:

- $h = h(n)$ is a sequence satisfying $\lim_{n \to \infty} h = 0$, $\lim_{n \to \infty} nh^{2\nu+1} = \infty$, $1 \le \nu$,
- $m \in C^{k_0}[0,1]$, $\nu + k \le k_0$,
- $K \in S_{0,k-\nu}^{\nu}$, i.e., $K^{(\nu)} \in S_{\nu,k}^{0}$ (see Granovsky and Müller (1991); Marron and Nolan (1988)),
- the point x is an inner point of the interval $[0,1]$, i.e., there exists an index n_0 such that
$$h < x < 1 - h, \ \forall n \ge n_0.$$

Gasser – Müller estimators (6.5) are suitable for the estimation of the ν-th derivative of m. The estimate is defined, at a given point x, by

$$\widehat{m}^{(\nu)}(x,h) = \frac{1}{h^{\nu}} \sum_{i=1}^{n} Y_i \int_{s_{i-1}}^{s_i} K_h(t-x)\,dt, \tag{6.36}$$

where

$$s_i = \frac{x_i + x_{i+1}}{2} = \frac{2i + 1}{2n}, \quad i = 1, \ldots, n - 1, \quad s_0 = 0, \quad s_n = 1.$$

It was shown, *e.g.*, in Müller (1988) that (6.36) is a consistent estimator of $m^{(\nu)}(x)$ under the assumptions given above.

We again consider MISE as a criterion of the quality of the estimate

$$\text{MISE}\{\widehat{m}^{(\nu)}(\cdot, h)\} = E \int \{\widehat{m}^{(\nu)}(x, h) - m^{(\nu)}(x)\}^2 dx.$$

It can be derived straightforwardly that

$$\text{MISE}\{\widehat{m}^{(\nu)}(\cdot, h)\} = \frac{\sigma^2 V(K^{(\nu)})}{nh^{2\nu+1}} + h^{2(k-\nu)} \frac{\beta_k^2(K^{(\nu)})}{(k!)^2} A_k$$
$$+ o\left\{ h^{2(k-\nu)} + (nh^{2\nu+1})^{-1} \right\},$$

where $\beta_k(K^{(\nu)}) = \int\limits_{-1}^{1} x^k K^{(\nu)}(x) dx$.

Obviously, AMISE takes the form

$$\text{AMISE}\{\widehat{m}^{(\nu)}(\cdot, h)\} = \frac{\sigma^2 V(K^{(\nu)})}{nh^{2\nu+1}} + h^{2(k-\nu)} \frac{\beta_k^2(K^{(\nu)})}{(k!)^2} A_k \qquad (6.37)$$

and further the optimal bandwidth minimizing AMISE

$$h_{opt,\nu,k}^{2k+1} = \frac{\sigma^2 (2\nu + 1)(k!)^2}{2(k - \nu)nA_k} \gamma_{\nu,k}^{2k+1}, \qquad (6.38)$$

where

$$\gamma_{\nu,k}^{2k+1} = \frac{V(K^{(\nu)})}{\beta_k^2(K^{(\nu)})}.$$

A similar calculation as in the density derivative estimating leads to the expression for $\text{AMISE}\{\widehat{m}^{(\nu)}(\cdot, h_{opt,\nu,k})\}$

$$\text{AMISE}\{\widehat{m}^{(\nu)}(\cdot, h_{opt,\nu,k})\} = T(K^{(\nu)}) \frac{\sigma^2 (2k + 1)\gamma_{\nu,k}^{2\nu+1}}{2n(k - \nu)h_{opt,\nu,k}^{2\nu+1}}, \qquad (6.39)$$

where

$$T(K^{(\nu)}) = \left(|\beta_k|^{2\nu+1} V(K^{(\nu)})^{k-\nu} \right)^{\frac{2}{2k+1}}.$$

This formula is very useful for the factor method. We can use similar ideas as in the case of density derivative estimating, see formulas (2.33) and (2.34). To complete all considerations about this procedure, it is necessary to mention the method for estimating the optimal bandwidth for the first derivative, which is described in the following paragraph.

6.4.1 *Choosing the bandwidth*

We will follow the estimator of the optimal bandwidth for the first derivative estimation. One of the most popular methods is the modified cross-validation proposed, *e.g.*, in Müller (1988); Zelinka and Horová (2001). The objective function is defined as

$$\mathrm{CV}_{(1)}(h) = \frac{1}{n-1} \sum_{i=1}^{n-1} \left\{ \frac{Y_{i+1} - Y_i}{x_{i+1} - x_i} - \widehat{m}^{(1)}_{-(i,i+1)}(c_i, h) \right\},$$

where $c_i = (x_i + x_{i+1})/2$, $i = 1, \ldots, n-1$, and $\widehat{m}^{(1)}_{-(i,i+1)}(c_i, h)$ stands for the Gasser – Müller estimate of $m^{(1)}$ at the point c_i based on the data pairs (x_j, Y_j), $j \in \{1, \ldots, n\}$ with points (x_i, Y_i) and (x_{i+1}, Y_{i+1}) deleted. The value $h_{opt,1,k}$ which minimizes $\mathrm{CV}_{(1)}(h)$, *i.e.*,

$$\hat{h}_{opt,1,k} = \arg\min_{h \in H_n} \mathrm{CV}_{(1)}(h),$$

is defined as the estimate of the optimal bandwidth $h_{opt,1,k}$.

6.5 Automatic procedure for simultaneous choice of the kernel, the bandwidth and the kernel order

As well as in density estimation it makes sense to propose an automatic procedure for kernel regression which selects simultaneously the optimal kernel, the bandwidth and the kernel order.

In connection with (6.39) we define

$$L(k) = T(K_{opt}^{(\nu)}) \frac{(2k+1)\gamma_{\nu,k}^{2\nu+1}}{2n(k-\nu)h_{opt,\nu,k}^{2\nu+1}} \tag{6.40}$$

as a kernel order selection criterion which is needed to minimize with respect to k.

Let us denote

$$I_\nu(k_0) = \left\{ \nu + 2j, \ j = 0, \ldots, \left[\frac{k_0 - \nu}{2} \right] \right\}$$

the set of all values of kernel order which will be taken in account.

By combining the ideas developed in the previous paragraphs we summarize main steps of the algorithm.

We present the special case when $\nu = 0$ and k even:

<u>Step 1.</u> For any $k \in I_0(k_0)$ find the optimal kernel $K_{opt} \in S_{0,k}^0$ given by the formula in Theorem 1.2 or in Table 1.1 and compute the canonical factor γ_{0k}.

Step 2. For any $k \in I_0(k_0)$ and $K_{opt} \in S^0_{0,k}$ find the estimate of the optimal
bandwidth $\hat{h}_{opt,0,k}$ by means of one of the methods given in Sec. 6.3.

Step 3. For any $k \in I_0(k_0)$ compute the selection criterion $L(k)$ in which
the values of K_{opt} and $\hat{h}_{opt,0,k}$ are those obtained at Steps 1 and 2.

Step 4. Compute the optimal order $\hat{k}$ carrying out the minimization
of (6.40) for $\nu = 0$.

Step 5. Use the parameters selected in the previous steps to get the optimal
kernel estimate of m, *i.e.*,

$$\widehat{m}(x, \hat{h}_{opt,0,\hat{k}}) = \sum_{i=1}^{n} W_i(x, \hat{h}_{opt,0,\hat{k}}) Y_i. \tag{6.41}$$

The procedure for estimating $m^{(\nu)}$ can be realized with the use of the factor
method for ν even. In the case of odd ν, it is necessary to estimate $\hat{h}_{opt,1,k}$
and then use the factor method.

This algorithm is also implemented in the MATLAB toolbox, for more
see Sec. 6.9, §6.9.4.

6.6   Boundary effects

If the support of the true regression curve is bounded then most nonparametric methods give estimates that are severely biased in regions near
the endpoints. To be specific, the bias of $\widehat{m}(x)$ is of order $O(h)$ rather
than $O(h^2)$ for $x \in [0, h] \cup [1 - h, 1]$. This boundary problem affects
the global performance visually and also in terms of a slower rate of convergence in the usual asymptotic analysis. It has been recognized as a serious
problem and many works are devoted to reducing the effects.

Gasser and Müller (1979); Gasser *et al.* (1985); Granovsky and Müller
(1991) and Müller (1991) discuss boundary kernel methods. These methods
are also implemented in the MATLAB toolbox, for more see Sec. 6.9.

Another approach to the boundary problems are reflection methods
which generally consist in reflecting data about the boundary points and
then estimating regression function. These methods are discussed, *e.g.*,
in Schuster (1985); Hall and Wehrly (1991). Reflection principles used
in kernel density estimation can be also adapted to kernel regression, see
some references in Sec. 2.7. The regression estimator with the assumption
of the "cyclic" model described in §6.3.4 can be also considered as the special case of a reflection technique. A short comparative study of methods
for boundary effects eliminating is given in Koláček and Poměnková (2006).

6.7 Simulations

We carry out two simulation studies to compare the performance of the bandwidth estimates. The comparison is done by the following way. The observations, Y_i, for $i = 1, \ldots, n = 74$, are obtained by adding independent Gaussian random variables with mean zero and variance $\sigma^2 = 0.2$ to some known regression function. Both regression functions used in our simulations are illustrated in Fig. 6.10 and Fig. 6.11, respectively. They are not chosen randomly for our comparison. The first one is suitable for the extension to the cyclic model, on the other side, the second function does not satisfy the assumption for the cyclic model.

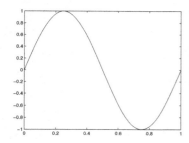   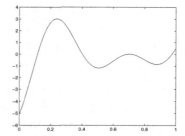

Fig. 6.10 $m(x) = \sin(2\pi x)$.   Fig. 6.11 $m(x) = -6\,\frac{\sin(11\,x+5)}{\cot(x-7)}$.

Two hundred series are generated. For each data set, we estimate the optimal bandwidth by all mentioned methods, *i.e.*, for each method we obtain 200 estimates. Since we know the optimal bandwidth, we compare it with the mean of estimates and look at their standard deviation, which describes the variability of all methods. The Nadaraya – Watson estimator is used in all cases.

In our simulations, we compare the following six methods for optimal bandwidth estimation listed in Table 6.1.

Means and standard deviations for the above mentioned methods are listed in tables. The histograms demonstrate the distribution of all 200 estimates of optimal bandwidth for all methods. The vertical dashed line represents the value of the theoretical optimal bandwidth $h_{opt,0,k}$. Kernels used in our simulation study are listed in Table 6.2.

Table 6.1 Methods included to the simulation study.

Method	Notation
Cross-validation	CV
Rice's bandwidth selector	Rice-pen
ET bandwidth selector	ET-pen
Mallows' method	Mallows
Minimization of $\widetilde{R}_n(h)$	Fourier
Plug-in method	plug-in

Table 6.2 Kernels of class $S_{0,k}$ used in simulations.

k	$K(x)$ on $[-1, 1]$
2	$-\frac{3}{4}(x^2 - 1)$
4	$\frac{15}{32}(x^2 - 1)(7x^2 - 3)$
6	$-\frac{105}{256}(x^2 - 1)(33x^4 - 30x^2 + 5)$

More comparisons and simulations can be found in Koláček (2005).

6.7.1 *Simulation 1*

In this case, we used the regression function

$$m(x) = \sin(2\pi x).$$

Table 6.3 summarizes the sample means and the sample standard deviations of bandwidth estimates, $E(\hat{h})$ is the average of all 200 values and $std(\hat{h})$ is their standard deviation. Figure 6.12 illustrates the histogram of results of all 200 experiments for $k = 2$.

As we see, the standard deviation of all results obtained by the last two methods, *i.e.*, Fourier and plug-in method, is less than the value for the case of others methods and also the mean of these results is closer to the theoretical optimal bandwidth. The reason is that the regression function is smooth and satisfies the conditions for the extension to the cyclic design. Thus the last two methods work very well in this case.

Table 6.3 Means and standard deviations of all estimates, $m(x) = \sin(2\pi x)$.

| | $k = 2$ | | $k = 4$ | | $k = 6$ | |
| | $h_{opt,0,2} = 0.1374$ | | $h_{opt,0,4} = 0.3521$ | | $h_{opt,0,6} = 0.5783$ | |
	$E(\hat{h})$	$std(\hat{h})$	$E(\hat{h})$	$std(\hat{h})$	$E(\hat{h})$	$std(\hat{h})$
CV	0.1063	0.0391	0.2232	0.0712	0.3273	0.1056
Rice-pen	0.1222	0.0329	0.2493	0.0585	0.3691	0.0877
ET-pen	0.1114	0.0342	0.2312	0.0625	0.3397	0.0915
Mallows	0.1269	0.0402	0.3354	0.0938	0.4432	0.1078
Fourier	0.1409	0.0095	0.3625	0.0306	0.4967	0.0172
plug-in	0.1383	0.0074	0.3422	0.0348	0.5604	0.0623

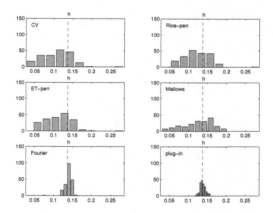

Fig. 6.12 Distribution of all 200 results for $k = 2$.

6.7.2 *Simulation 2*

In the second simulation, we generated data to the regression function

$$m(x) = -6\,\frac{\sin(11\,x + 5)}{\cot(x - 7)}.$$

Table 6.4 summarizes the sample means and the sample standard deviations of bandwidth estimates, $E(\hat{h})$ is the average of all 200 values and $std(\hat{h})$ is their standard deviation. Figure 6.13 illustrates the histogram of results of all 200 experiments for $k = 4$. The distribution of results for other cases ($k = 2$, $k = 6$) was similar.

Table 6.4 Means and standard deviations of all estimates, $m(x) = -6 \frac{\sin(11\,x+5)}{\cot(x-7)}$.

| | $k = 2$ | | $k = 4$ | | $k = 6$ | |
| | $h_{opt,0,2} = 0.0631$ | | $h_{opt,0,4} = 0.1495$ | | $h_{opt,0,6} = 0.2312$ | |
	$E(\hat{h})$	$std(\hat{h})$	$E(\hat{h})$	$std(\hat{h})$	$E(\hat{h})$	$std(\hat{h})$
CV	0.0483	0.0104	0.1111	0.0252	0.1652	0.0378
Rice-pen	0.0581	0.0064	0.1273	0.0186	0.1844	0.0309
ET-pen	0.0558	0.0062	0.1211	0.0191	0.1794	0.0311
Mallows	0.0292	0.0033	0.0584	0.0080	0.0882	0.0106
Fourier	0.0372	0.0052	0.0766	0.0135	0.1149	0.0221
plug-in	0.0673	0.0025	0.1429	0.0037	0.2233	0.0042

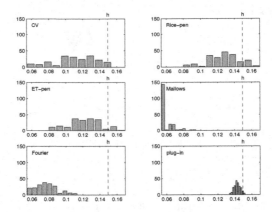

Fig. 6.13 Distribution of all 200 results for $k = 4$.

It is evident that the best results were obtained by the plug-in method. This method was successful despite the fact that the regression function does not meet assumptions for the extension to the cyclic model. The other methods often resulted in smaller bandwidths. The variance of these criteria was also significant, especially for the cross-validation method. Mallows' and Fourier methods gave the worst results. Both of them depend on the estimation of σ which was rather small in this case.

6.8 Application to real data

The main goal of this section is to make a comparison of all mentioned bandwidth estimators on a real data set. We used data from the Czech Hydrometeorological Institute and followed the average spring temperatures in Prague's Klementinum in 1771 – 2000, *i.e.*, the sample size was $n = 230$. We transformed data to the interval $[0, 1]$ and used all selectors considered in the previous section to get the optimal bandwidth. We used the Nadaraya – Watson estimator with the kernels listed in Table 6.2. All estimates of optimal bandwidth are listed in Table 6.5. The values can be divided into three groups. The first group corresponds with Mallows' method which gave the smallest bandwidths. As the second group we can denote the cross-validation and both penalizing methods, since all results were similar to each other. The third group is represented by the methods based on Fourier transformation. The application of these criteria leaded to the highest values of bandwidth estimates.

Table 6.5   Optimal bandwidth estimates for temperatures data.

	$k = 2$	$k = 4$	$k = 6$
CV	0.0743	0.2955	0.4011
Rice-pen	0.0743	0.2955	0.4042
ET-pen	0.0743	0.2955	0.4011
Mallows	0.0649	0.0923	0.1448
Fourier	0.1993	0.4549	0.5011
plug-in	0.1914	0.4233	0.6569

For kernel regression estimation we used the kernel $K \in S_{0,4}$ (see Table 6.2). In other cases, we obtained similar estimates. Figure 6.14 illustrates the kernel regression estimate with the smoothing parameter $\hat{h} = 0.0923$ which was obtained by Mallows' method.

Figure 6.15 shows the kernel regression estimate with the smoothing parameter $\hat{h} = 0.2955$. This value was found by cross-validation and both penalizing methods.

Figure 6.16 illustrates the quality of the kernel regression estimate with parameter $\hat{h} = 0.4233$ obtained by criteria based on Fourier transformation.

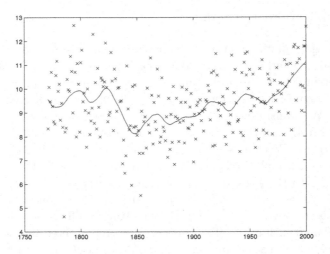

Fig. 6.14　Kernel regression estimate with $\hat{h} = 0.0923$.

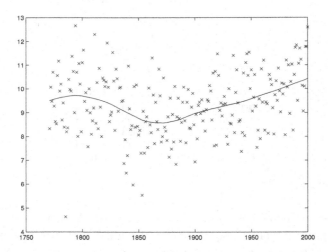

Fig. 6.15　Kernel regression estimate with $\hat{h} = 0.2955$.

Since we do not know the true regression function $m(x)$ it is hard to assess objectively which one of kernel estimates is the best. It is very important to realize the fact that the final decision about the estimate is partially subjective because the estimates of the bandwidth are only asymptotically optimal. The values summarized in the table and figures show that the es-

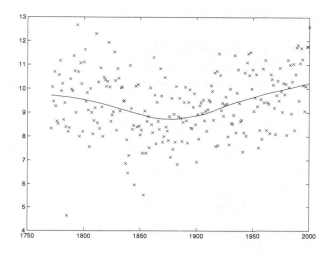

Fig. 6.16 Kernel regression estimate with $\hat{h} = 0.4233$.

timate with the smoothing parameter obtained by Mallows' criterion is undersmoothed. On the other hand, the methods based on Fourier transformation, *i.e.*, "Fourier" and "plug-in" tend to high values of the bandwidth and the estimate seems to be oversmoothed. In the context of these considerations, the estimate with parameters obtained by cross-validation and both penalizing methods appears to be sufficient.

6.9　Use of MATLAB toolbox

The toolbox can be downloaded from the web page
`http://www.math.muni.cz/english/science-and-research/`
`developed-software/232-matlab-toolbox.html`.

6.9.1　*Running the program*

The *Start menu* (Fig. 6.17) for kernel regression is called up by the command `ksregress`.

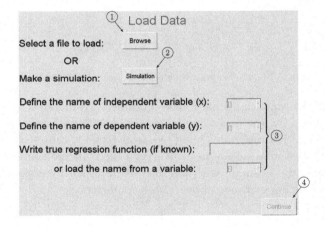

Fig. 6.17　Start menu.

You can skip this menu by typing input data as an argument `ksregress(x,y)`, where the vectors x and y should be of the same length n and they mark x and y axes of measurements. If we know also the true regression function $m(x)$ (for example for simulated data), we can set it as the next argument. For more see `help ksregress`. After executing this command directly the window in Fig. 6.19 is called up.

In the *Start menu*, you have several possibilities how to define input data. You can load it from a file (button ①) or simulate data (button ②). In the fields ③ you can list your variables in the current workspace to define input data. If your workspace is empty, these fields are nonactive. If you know the true regression function of the model, you can write it to the text field or load it from a variable. If you need to simulate a regression model, press button ②. Then the menu for simulation (Fig. 6.18) is called up.

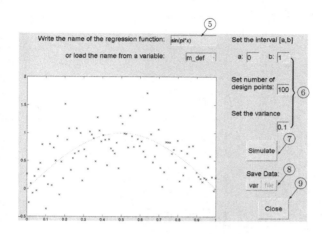

Fig. 6.18   Simulation menu.

In the *Simulation menu*, first, set the regression function. You can write it to the text field ⑤ (after doing it press ENTER or click anywhere outside the field) or load it from a variable. In the fields ⑥ specify the interval, the number of design points and the variance. You can simulate a regression model by pressing button ⑦. Then you can save data to variables and then as a file by using buttons ⑧. If you have done the simulation, press button ⑨. The *Simulation menu* will be closed and you will be returned to the *Start menu*. In this menu, you can redefine the input data. If you want to continue, press button ④. The menu will be closed and the *Basic menu* (see Fig. 6.19) will be called up.

6.9.2   *Main figure*

This menu (Fig. 6.19) was called up from the *Start menu* or directly from the command line (see **help ksregress**). The values of independent variable x are automatically transformed to the interval $[0, 1]$. Symbols × mark the measurements after this transformation. If you want to show the original data, use button ⑪. Button ⑫ ends the application. Press button ⑩ to continue. Other buttons are nonactive.

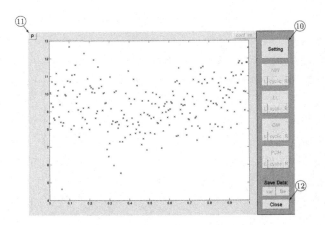

Fig. 6.19   Basic menu.

6.9.3 *Setting the parameters*

Button $\circled{10}$ calls up the menu for setting of the parameters which will be used for kernel regression, see Fig. 6.20.

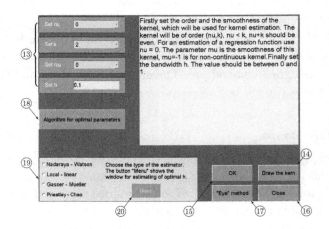

Fig. 6.20   Setting of parameters.

In the array ⑬, we can set kernel regression parameters. First, set the order of the kernel (ν, k), where $\nu + k$ should be even, for the regression estimation $\nu = 0$ is used. The parameter μ is the smoothness of this kernel. If you want to draw the kernel, use button ⑭. Finally set the bandwidth h. The value should be between 0 and 1. To confirm the setting use ⑮, to close the window, use ⑯. The other buttons are useful for choosing the optimal bandwidth.

Button ⑰ calls up the *"Eye-control" menu* (see Fig. 6.22), where we can change the value of h and observe the effect upon the final estimate. Button ⑱ starts the algorithm for estimation of optimal kernel order and optimal bandwidth (see §6.9.4). This algorithm automatically sets the values of optimal parameters in the array ⑬. By selecting one type of kernel estimators in ⑲ you make active ⑳. This button calls up the menu for using and comparing various methods for choosing the optimal smoothing parameter h (see §6.9.6).

6.9.4 *Estimation of optimal parameters*

Button ⑱ calls up an algorithm for the estimation of optimal order of kernel and optimal bandwidth. At first, it is necessary to set the type of kernel estimator:

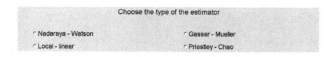

Next, the menu for estimating the optimal bandwidth is called up (see Fig. 6.21).

By choosing one of the methods in the array ㉑ we make active button ㉓ which starts the computation of optimal parameters k and h (we suppose $K \in S_{0,k}^0$). In the array ㉒, we can set limits for the parameter k, the default values are $k_{min} = 2$, $k_{max} = 12$. The results of the computation are automatically set in the array ⑬.

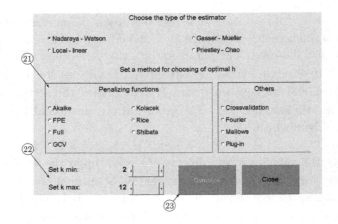

Fig. 6.21   Estimation of optimal parameters.

6.9.5   *Eye-control method*

Button ⑰ calls up a window, where we can change the value of parameter h and observe the effect of these changes upon the final estimate (see Fig. 6.22).

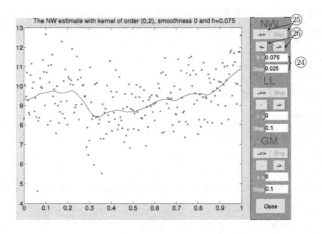

Fig. 6.22   "Eye-control" menu.

In the arrays ㉔, set the starting value of parameter h and the step (it can be positive or negative number) for the size of changes of h. The left button $\textcircled{25}$ starts a sequence of pictures representing the quality of estimation dependence on h. The right button stops the sequence. You can change the value of h only one step more or less by buttons $\textcircled{26}$.

6.9.6 *Comparing of methods for bandwidth selection*

Button $\textcircled{20}$ calls up the window for using and comparing of various methods for the optimal bandwidth selection (see Fig. 6.23).

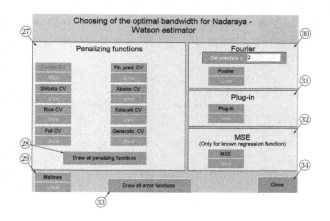

Fig. 6.23 Bandwidth estimators.

In this window, all mentioned bandwidth estimators are summarized. In the array $\textcircled{27}$, there are presented:

- method of penalizing functions – see §6.3.3
- the cross-validation method (denoted as "Classic CV") – see §6.3.2.

By clicking on the button with the method's name, the optimal bandwidth is computed by the actual method. The button "draw" calls up a graph of the minimized error function. To draw all penalizing functions applied up to this time use $\textcircled{28}$. Button $\textcircled{29}$ represents Mallows' method (see §6.3.1), button $\textcircled{30}$ denotes the method of Fourier transformation described in the first part of §6.3.4 and $\textcircled{31}$ marks the plug-in method, *i.e.*, the bandwidth estimate $\hat{h}_{\mathrm{PI}}$ given by (6.35). If the original regression function is known (for example for simulated data), we can compute the value of optimal bandwidth as a minimum of AMISE$\{\widehat{m}(\cdot, h)\}$ (see the formula

(6.12)) in the array ㉜. To do this computation, it is necessary to have the *Symbolic toolbox* installed on your computer. If this toolbox is not installed or the regression function is not defined on input, the array is not active. For the graph of all error functions and their minimal values use ㉝. Button ㉞ closes the application.

6.9.7   *The final estimation*

If you have set all values in the window for parameters setting (see Fig. 6.20) and if you want to go to the final estimation of the regression function, confirm your setting by button ⑮. It calls up the *Basic menu*, where all buttons are active already.

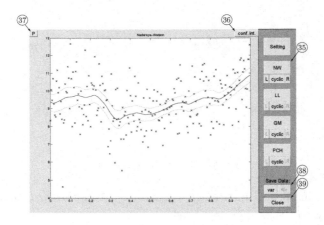

Fig. 6.24   Original data and the final kernel regression estimate.

By clicking on the button with the estimator's name (for example ㉟ for the Nadaraya–Watson estimator), the relevant regression estimate is drawn (solid line in the figure). The button "cyclic" shows the regression estimate with using the assumption of cyclic model. By using buttons "L" and "R" we get the estimate near the boundary of the interval obtained by using special boundary kernels (L=left, R=right). Button ㊱ draws confidence intervals (dashed) computed by using the formula (6.14). To do this computation, it is necessary to have the *Stats toolbox* installed on your computer. If this toolbox is not installed, the button is not active. Button ㊲ shows original data and the final estimate. You can also save data to variables and then as a file by using buttons ③⑧. Button ③⑨ ends the application.

6.10 Complements

Proof of Lemma 6.1.

Proof.

$$\widetilde{R}_n(h) = \frac{4\pi}{n} \sum_{j=2}^{N} \widetilde{I}_{Y_j} \{1 - w_j^-\}^2 - \hat{\sigma}^2 + 2\hat{\sigma}^2 w_1$$

$$= \frac{4\pi}{n} \sum_{j=2}^{J_1-1} I_{Y_j} \{1 - w_j^-\}^2 + \frac{4\pi}{n} \sum_{j=J_1}^{N} \frac{\hat{\sigma}^2}{2\pi} \{1 - w_j^-\}^2 - \hat{\sigma}^2 + 2\hat{\sigma}^2 w_1$$

$$= \frac{4\pi}{n} \sum_{j=2}^{J_1-1} \left\{ I_{Y_j} - \frac{\hat{\sigma}^2}{2\pi} \right\} \{1 - w_j^-\}^2 + \frac{\hat{\sigma}^2}{n} \sum_{j=1}^{n} \{1 - w_j^-\}^2 - \hat{\sigma}^2 + 2\hat{\sigma}^2 w_1$$

$$= \frac{4\pi}{n} \sum_{j=2}^{J_1-1} \left\{ I_{Y_j} - \frac{\hat{\sigma}^2}{2\pi} \right\} \{1 - w_j^-\}^2 + \frac{\hat{\sigma}^2}{n} \left(n - 2n w_1 + \sum_{j=1}^{n} (w_j^-)^2 \right)$$
$$\quad - \hat{\sigma}^2 + 2\hat{\sigma}^2 w_1$$

$$= \frac{4\pi}{n} \sum_{j=2}^{J_1-1} \left\{ I_{Y_j} - \frac{\hat{\sigma}^2}{2\pi} \right\} \{1 - w_j^-\}^2 + \frac{\hat{\sigma}^2}{n} \sum_{j=1}^{n} (w_j^-)^2. \qquad \square$$

Proof of Theorem 6.5.

Proof.

$$\sum_{j=1}^{n} (w_j^-)^2 = \sum_{j=1}^{n} |w_j^-|^2 = \sum_{j=1}^{n} w_j^- \overline{w_j^-}$$

$$= \sum_{j=1}^{n} \sum_{s=-n+1}^{n-1} \sum_{t=-n+1}^{n-1} W_0(x_s, h) W_0(x_t, h) e^{\frac{i2\pi(t-s)(j-1)}{n}}$$

$$= \sum_{s=-n+1}^{n-1} \sum_{t=-n+1}^{n-1} W_0(x_s, h) W_0(x_t, h) \sum_{j=0}^{n-1} e^{\frac{i2\pi(t-s)j}{n}}.$$

Let us compute the last sum

$$\sum_{j=0}^{n-1} e^{\frac{i2\pi(t-s)j}{n}} = \begin{cases} \dfrac{e^{i2\pi(t-s)}-1}{e^{\frac{i2\pi(t-s)}{n}}-1} = 0, & \text{for } s \neq t, \\[2mm] n, & \text{for } s = t. \end{cases}$$

Thus we obtain

$$\sum_{j=1}^{n}(w_j^-)^2 = n \sum_{s=-n+1}^{n-1} W_0^2(x_s, h). \qquad (6.42)$$

Let us recall an interesting formula which can be found, *e.g.*, in Wand and Jones (1995)

$$\frac{1}{n}\sum_{i=1}^{n} K_h(x_i - x) = \int_{-1}^{1} K(u)du + o(1) = 1 + o(1)$$

for $h < x < 1 - h$.

By using similar ideas on the cyclic model we can arrive at the formula for $x = 0$

$$\frac{1}{n}\sum_{i=-n+1}^{n-1} K_h(x_i) = 1 + o(1). \qquad (6.43)$$

Similarly, it also holds

$$\frac{1}{n}\sum_{i=-n+1}^{n-1} K_h^2(x_i) = \frac{1}{h}\int_{-1}^{1} K^2(u)du + o(h^{-1}). \qquad (6.44)$$

If we put (6.43) into the formula (6.24), we arrive at the following expression

$$W_0(x_j, h) = \frac{1}{n}K_h(x_j) + o(1), \quad j = 1, \ldots, n. \qquad (6.45)$$

Now, we employ this expression in (6.42) and then use (6.44)

$$\sum_{j=1}^{n}(w_j^-)^2 = \sum_{s=-n+1}^{n-1} \frac{1}{n}K_h^2(x_s) + o(1)$$

$$= \frac{1}{h}\int_{-1}^{1} K^2(x)dx + o(h^{-1}).$$

$\square$

Proof of Theorem 6.6.

Proof. It is an easy exercise to express w_j^- as

$$w_j^- = \sum_{s=-n+1}^{n-1} W_0(x_s, h)\cos\left(\frac{2\pi s(j-1)}{n}\right), \quad j = 1, \ldots, n.$$

We use similar ideas as in the previous proof.

$$\frac{1}{(2\pi j)^k}(1 - w_{j+1}^-) = \frac{1}{(2\pi j)^k}\left\{1 - \sum_{s=-n+1}^{n-1} W_0(x_s, h)\cos\left(\frac{2\pi sj}{n}\right)\right\}$$

$$= \frac{1}{(2\pi j)^k}\left\{1 - \frac{1}{n}\sum_{s=-n+1}^{n-1} K_h(x_s)\cos\left(\frac{2\pi sj}{n}\right)\right\} + o(1).$$

By using the same technique as in the proof of Theorem 6.5 we can show that

$$\frac{1}{n}\sum_{s=-n+1}^{n-1} K_h(x_s)\cos\left(\frac{2\pi sj}{n}\right) = \int_{-1}^{1} K(u)\cos(2\pi ju)du + o(1).$$

Thus we get

$$\frac{1}{(2\pi j)^k}(1 - w_{j+1}^-) = \frac{1}{(2\pi j)^k}\left\{1 - \int_{-1}^{1} K_h(u)\cos(2\pi ju)du\right\} + o(1)$$

$$= \frac{1}{(2\pi j)^k}\int_{-1}^{1}\{1 - \cos(2\pi ju)\}K_h(u)du + o(1).$$

We replace the function $1 - \cos(2\pi ju)$ by Taylor's polynomial of degree k. Let R_k is an error of this approximation

$$\frac{1}{(2\pi j)^k}(1 - w_{j+1}^-) = \frac{1}{(2\pi j)^k}\int_{-1}^{1}\left\{\frac{(2\pi ju)^2}{2} - \frac{(2\pi ju)^4}{24} + \dots\right.$$

$$\left.\dots + \frac{(-1)^{\frac{k}{2}+1}(2\pi ju)^k}{k!}\right\}K_h(u)du + \frac{R_k}{(2\pi j)^k} + o(h^k)$$

$$= \frac{(-1)^{\frac{k}{2}+1}}{k!}\int_{-1}^{1} u^k K_h(u)du + \frac{R_k}{(2\pi j)^k} + o(h^k)$$

$$= (-1)^{\frac{k}{2}+1}\frac{h^k}{k!}\int_{-1}^{1} x^k K(x)dx + \frac{R_k}{(2\pi j)^k} + o(h^k).$$

The assumptions for the index j, $j < J$, yield $\left|\frac{R_k}{(2\pi j)^k}\right| \leq \frac{\varepsilon}{(2\pi)^k}$ for any $\varepsilon > 0$.

$\square$

The main idea of derivation of the estimate $\widehat{A}_k$:

We consider the error function $\widetilde{R}_n(h)$ as the estimate of

$$\text{AMISE}\{\widehat{m}(\cdot, h)\} = \underbrace{\frac{V(K)\sigma^2}{nh}}_{\text{AIV}} + \underbrace{\left(\frac{\beta_k}{k!}\right)^2 A_k h^{2k}}_{\text{AISB}}.$$

According to Lemma 6.1, it takes the form

$$\widetilde{R}_n(h) = \frac{\hat{\sigma}^2}{n} \sum_{j=1}^{n} (w_j^-)^2 + \frac{4\pi}{n} \sum_{j=2}^{J_1-1} \left\{ I_{Y_j} - \frac{\hat{\sigma}^2}{2\pi} \right\} \{1 - w_j^-\}^2.$$

Theorem 6.5 says that the first term of $\widetilde{R}_n(h)$ estimates the AIV, *i.e.*,

$$\frac{\hat{\sigma}^2}{n} \sum_{j=1}^{n} (w_j^-)^2 = \frac{V(K)\sigma^2}{nh} + o\left\{(nh)^{-1}\right\}.$$

Thus the second term estimates the AISB. Together with this idea we use the result of Theorem 6.6

$$\frac{1}{(2\pi j)^k}(1 - w_{j+1}^-) = (-1)^{\frac{k}{2}+1} \frac{h^k}{k!} \beta_k + \xi + o\left(h^k\right).$$

Let us express the second term of $\widetilde{R}_n(h)$

$$\frac{4\pi}{n} \sum_{j=1}^{J-2} \left\{ I_{Y_{j+1}} - \frac{\hat{\sigma}^2}{2\pi} \right\} \{1 - w_{j+1}^-\}^2$$

$$= \frac{4\pi}{n} \sum_{j=1}^{J-2} \frac{1}{(2\pi j)^{2k}} \{1 - w_{j+1}^-\}^2 (2\pi j)^{2k} \left\{ I_{Y_{j+1}} - \frac{\hat{\sigma}^2}{2\pi} \right\}$$

$$= \left(\frac{\beta_k}{k!}\right)^2 h^{2k} \frac{4\pi}{n} \sum_{j=1}^{J-2} (2\pi j)^{2k} \left\{ I_{Y_{j+1}} - \frac{\hat{\sigma}^2}{2\pi} \right\} + O(\xi/n) + o(h^k).$$

This term is an approximation of AISB, thus we obtain $\widehat{A}_k$.

Chapter 7

Multivariate kernel density estimation

The present chapter is devoted to the extension of the univariate kernel density estimate to the multivariate setting. This extension is not without any problem. The most general smoothing parameterization of the kernel estimator in d-dimensions requires the specification of entries of a positive definite bandwidth matrix. The bandwidth matrix controls both the smoothness and the orientation of the multivariate smoothing. The multivariate kernel density estimator we are going to deal with is a direct extension of the univariate estimator (see, *e.g.*, Wand and Jones (1995)). Papers Chacón *et al.* (2011); Duong *et al.* (2008) also investigated general kernel estimators of multivariate density derivative using general (or unconstrained) bandwidth matrix selectors. In this chapter we focus on estimates of both multivariate density and its gradient.

7.1 Basic definition

Let a d-variate random sample $\mathbf{X}_1, \ldots, \mathbf{X}_n$ be drawn from a density f. The kernel density estimator $\hat{f}$, for $\mathbf{x} \in \mathbb{R}^d$, is defined as

$$\hat{f}(\mathbf{x}, \mathbf{H}) = \frac{1}{n} \sum_{i=1}^{n} K_{\mathbf{H}}(\mathbf{x} - \mathbf{X}_i), \tag{7.1}$$

where K is a kernel function, which is often taken to be a d-variate symmetric probability function, $\mathbf{H}$ is a positive definite $d \times d$ symmetric matrix and $K_{\mathbf{H}}$ is the scaled kernel function

$$K_{\mathbf{H}}(\mathbf{x}) = |\mathbf{H}|^{-1/2} K(\mathbf{H}^{-1/2}\mathbf{x})$$

with $|\mathbf{H}|$ the determinant of the matrix $\mathbf{H}$. This matrix is called a *bandwidth matrix*.

There are two types of multivariate kernels created from a symmetric univariate kernel k – a *product kernel* K^P, and a *spherically symmetric kernel* K^S:

$$K^P(\mathbf{x}) = \prod_{i=1}^{d} k(x_i), \qquad K^S(\mathbf{x}) = c_k k\left(\sqrt{\mathbf{x}^T \mathbf{x}}\right),$$

where $c_k^{-1} = \int k\left(\sqrt{\mathbf{x}^T \mathbf{x}}\right) d\mathbf{x}$. The choice of a kernel does not influence the estimate as significantly as the bandwidth matrix.

Example 7.1. Bivariate kernels for $k(x) = -\frac{3}{4}(1 - x^2)I_{[-1,1]}(x)$ (the Epanechnikov kernel) are presented in Fig. 7.1 and Fig. 7.2.

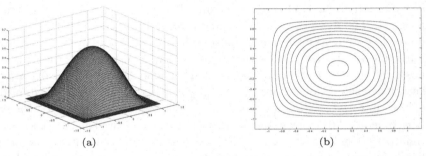

(a)   (b)

Fig. 7.1   Product kernel K^P, (a) 3-D plot, (b) contour plots.

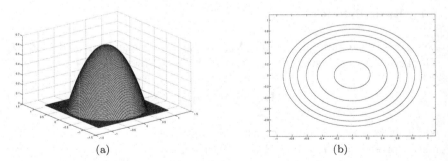

(a)   (b)

Fig. 7.2   Spherically symmetric kernel K^S, (a) 3-D plot, (b) contour plots.

The problem of the bandwidth matrix selection can be simplified by imposing constraints on $\mathbf{H}$. Let $\mathcal{H}$ denote a class of $d \times d$ positive definite

symmetric matrices. The matrix $\mathbf{H} \in \mathcal{H}$ (called a full matrix) has $d(d+1)/2$
entries which have to be chosen. A simplification can be obtained by im-
posing the restriction $\mathbf{H} \in \mathcal{H}_\mathcal{D}$, where $\mathcal{H}_\mathcal{D} \subset \mathcal{H}$ is a subclass of diagonal
positive definite matrices; $\mathbf{H} = diag(h_1^2, \ldots, h_d^2)$. A further simplification
follows from the restriction $\mathbf{H} \in \mathcal{H}_\mathcal{S}$ where $\mathcal{H}_\mathcal{S} = \{h^2 \mathbf{I}_d, \ h > 0\}$, $\mathbf{I}_d$ is $d \times d$
identity matrix and leads to the single bandwidth estimator (Wand and
Jones (1995)). Using the single bandwidth matrix parameterization class
$\mathcal{H}_\mathcal{S}$ is not advised for data which have different dispersions in co-ordinate
directions (Wand and Jones (1993)). On the other hand, the bandwidth
matrix selectors in the general $\mathcal{H}$ class are able to handle differently dis-
persed data, but it is not easy to determine the optimal bandwidth matrix
in this case. Using the diagonal matrix class $\mathcal{H}_\mathcal{D}$ seems to be a compromise
and this approach is often used in practice. These ideas can be easily seen
in the case of bivariate data, see Fig. 7.3.

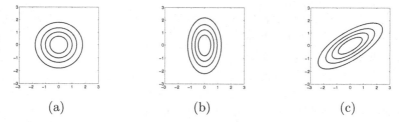

$$\qquad\quad \text{(a)} \qquad\qquad\qquad\quad \text{(b)} \qquad\qquad\qquad\quad \text{(c)}$$

Fig. 7.3   Contour plots of kernels parameterized by (a) $\mathbf{H} \in \mathcal{H}_\mathcal{S}$, (b) $\mathbf{H} \in \mathcal{H}_\mathcal{D}$, (c)
$\mathbf{H} \in \mathcal{H}$.

The kernel estimator of the gradient Df, i.e., the estimator of the column vector of the first partial derivatives of f is defined as (see Duong *et al.*
(2008))

$$\widehat{Df}(\mathbf{x}, \mathbf{H}) = \frac{1}{n} \sum_{i=1}^{n} DK_\mathbf{H}(\mathbf{x} - \mathbf{X}_i), \qquad (7.2)$$

where $DK_\mathbf{H}(\mathbf{x}) = |\mathbf{H}|^{-1/2} \mathbf{H}^{-1/2} DK(\mathbf{H}^{-1/2}\mathbf{x})$ and DK is the column vec-
tor of the partial derivatives of K.

Since we aim to investigate both density itself and its gradient in a similar way, we introduce the notation

$$\widehat{D^r f}(\mathbf{x}, \mathbf{H}) = \frac{1}{n} \sum_{i=1}^{n} D^r K_\mathbf{H}(\mathbf{x} - \mathbf{X}_i), \ r = 0, 1, \qquad (7.3)$$

where $D^0 f = f$, $D^1 f = Df$.

We make some additional assumptions and notations:

(A_1) The kernel function K satisfies the moment conditions $\int K(\mathbf{x})d\mathbf{x} = 1$, $\int \mathbf{x}K(\mathbf{x})d\mathbf{x} = \mathbf{0}$, $\int \mathbf{x}\mathbf{x}^T K(\mathbf{x})d\mathbf{x} = \beta_2 \mathbf{I}_d$, $\mathbf{I}_d$ is the $d \times d$ identity matrix.

(A_2) $\mathbf{H} = \mathbf{H}_n$ is a sequence of bandwidth matrices such that $n^{-1/2}|\mathbf{H}|^{-1/2}(\mathbf{H}^{-1})^r$, $r = 0, 1$ and the entries of $\mathbf{H}$ approach zero $((\mathbf{H}^{-1})^0$ is considered as equal to 1).

(A_3) Each partial density derivative of order $r + 2$, $r = 0, 1$, is continuous and square integrable.

(N_1) Let us denote $V(\rho) = \int\limits_{\mathbb{R}} \rho^2(x)dx$ for any square integrable scalar valued function ρ.

(N_2) Let us denote $V(g) = \int\limits_{\mathbb{R}^d} g(\mathbf{x})g^T(\mathbf{x})d\mathbf{x}$ for any square integrable vector valued function g.

(N_3) The matrix $DD^T = D^2$ is a Hessian matrix of a density f. Expressions like $DD^T = D^2$ involves "multiplications" of differentials in the sense that

$$\frac{\partial}{\partial x_i}\frac{\partial}{\partial x_j} = \frac{\partial^2}{\partial x_i \partial x_j}.$$

This means that $(D^2)^m$, $m \in \mathbb{N}$, is a matrix of the $2m$-th order partial differential operators.

(N_4) The vector $\text{vec}\mathbf{H}$ is $d^2 \times 1$ vector obtained by stacking columns of $\mathbf{H}$.

(N_5) Let $d^* = d(d+1)/2$, $\text{vech}\mathbf{H}$ is $d^* \times 1$ vector-half obtained from $\text{vec}\mathbf{H}$ by eliminating each of the above diagonal entries.

(N_6) The matrix $\mathbf{D}_d$ of size $d^2 \times d^*$ of ones and zeros such that

$$\mathbf{D}_d\text{vech}\mathbf{H} = \text{vec}\mathbf{H}$$

is called the *duplication matrix* of order d.

(N_7) The symbol $\mathbf{J}_d$ denotes $d \times d$ matrix of ones.

(N_8) In the rest of the text, $\int$ stands for $\int\limits_{\mathbb{R}^d}$ unless it is stated otherwise.

7.2 Statistical properties of the estimate

The quality of the estimate $\widehat{D^r f}$ can be expressed in terms of the Mean Integrated Square Error

$$\mathrm{MISE}_r\{\widehat{D^r f}(\cdot, \mathbf{H})\} = E \int ||\widehat{D^r f}(\mathbf{x}, \mathbf{H}) - D^r f(\mathbf{x})||^2 dx,$$

with $|| \cdot ||$ standing for the Euclidean norm, *i.e.*, $||\mathbf{v}||^2 = \mathbf{v}^T \mathbf{v} = \mathrm{tr}(\mathbf{v}\mathbf{v}^T)$. For the sake of simplicity we write the argument of MISE_r as $\mathbf{H}$. This error function can be also expressed as the standard decomposition

$$\mathrm{MISE}_r(\mathbf{H}) = \mathrm{IV}_r(\mathbf{H}) + \mathrm{ISB}_r(\mathbf{H}),$$

where $\mathrm{IV}_r(\mathbf{H}) = \int \mathrm{var}\{\widehat{D^r f}(\mathbf{x}, \mathbf{H})\} dx$ is the Integrated Variance and

$$\begin{aligned}
\mathrm{ISB}_r(\mathbf{H}) &= \int ||E\widehat{D^r f}(\mathbf{x}, \mathbf{H}) - D^r f(\mathbf{x})||^2 dx \\
&= \int \left|\left| \int K(\mathbf{z}) D^r f(\mathbf{x} - \mathbf{H}^{1/2}\mathbf{z}) d\mathbf{z} - D^r f(\mathbf{x}) \right|\right|^2 dx \\
&= \int ||(K_{\mathbf{H}} * D^r f)(\mathbf{x}) - D^r f(\mathbf{x})||^2 dx
\end{aligned}$$

is the Integrated Square Bias (the symbol $*$ denotes convolution).

From the same reason as earlier we again employ the Asymptotic Mean Integrated Square Error. The AMISE_r theorem has been proved, (*e.g.*, in Duong *et al.* (2008)) and reads as follows:

Theorem 7.1. *Let assumptions* $(A_1) - (A_3)$ *be satisfied. Then*

$$\mathrm{MISE}_r(\mathbf{H}) \simeq \mathrm{AMISE}_r(\mathbf{H}),$$

where

$$\mathrm{AMISE}_r(\mathbf{H}) = \underbrace{n^{-1}|\mathbf{H}|^{-1/2}\mathrm{tr}\left\{(\mathbf{H}^{-1})^r V(D^r K)\right\}}_{\mathrm{AIV}_r} + \underbrace{\frac{\beta_2^2}{4}\mathrm{vech}^T \mathbf{H}\boldsymbol{\Psi}_{4+2r}\mathrm{vech}\mathbf{H}}_{\mathrm{AISB}_r}.$$

$$(7.4)$$

The term $\boldsymbol{\Psi}_{4+2r}$ involves higher order derivatives of f and its subscript $4 + 2r$, $r = 0, 1$, indicates the order of derivatives used. It is a symmetric $d^* \times d^*$ matrix (see also Wand and Jones (1995) for the case $r = 0$).

7.2.1 *Exact MISE calculations*

Let $K = \phi_\mathbf{I}$ be the d-variate normal kernel and suppose that f is the normal mixture density $f(\mathbf{x}) = \sum_{l=1}^{k} w_l \phi_{\mathbf{\Sigma}_l}(\mathbf{x} - \boldsymbol{\mu}_l)$, where for each $l = 1, \ldots, k$, $\phi_{\mathbf{\Sigma}_l}$ is the d-variate $N(\mathbf{0}, \mathbf{\Sigma}_l)$ normal density and $\mathbf{w} = (w_1, \ldots, w_k)^T$ is a vector of positive numbers summing to one.

In this case, the exact formula for MISE_r was derived in Chacón *et al.* (2011). For $r = 0, 1$ it takes the form

$$\text{MISE}_r(\mathbf{H}) = 2^{-r} n^{-1} (4\pi)^{-d/2} |\mathbf{H}|^{-1/2} (\text{tr}\mathbf{H}^{-1})^r \tag{7.5}$$
$$+ \mathbf{w}^T \left\{ (1 - n^{-1})\mathbf{\Omega}_2 - 2\mathbf{\Omega}_1 + \mathbf{\Omega}_0 \right\} \mathbf{w},$$

where

$$(\mathbf{\Omega}_c)_{ij} = (-1)^r \phi_{c\mathbf{H}+\mathbf{\Sigma}_{ij}}(\boldsymbol{\mu}_{ij}) \left\{ \boldsymbol{\mu}_{ij}^T (c\mathbf{H} + \mathbf{\Sigma}_{ij})^{-2} \boldsymbol{\mu}_{ij} - 2\text{tr}(c\mathbf{H} + \mathbf{\Sigma}_{ij})^{-1} \right\}^r$$

with $\mathbf{\Sigma}_{ij} = \mathbf{\Sigma}_i + \mathbf{\Sigma}_j$, $\boldsymbol{\mu}_{ij} = \boldsymbol{\mu}_i - \boldsymbol{\mu}_j$.

7.3 Bandwidth matrix selection

The most important factor in multivariate kernel density estimates is a bandwidth matrix $\mathbf{H}$. Because of its role in controlling both the amount and the direction of smoothing this choice is particularly important. A common approach to the multivariate smoothing is to first rescale data so that the sample variances are equal in each dimension – this approach is called scalling or sphering data so the sample covariance matrix is the identity (see, *e.g.*, Duong (2007); Wand and Jones (1993)). But here we do not use this strategy.

Let $\mathbf{H}_{(\text{A})\text{MISE},r}$ stand for a bandwidth matrix minimizing $(\text{A})\text{MISE}_r$, *i.e.*,

$$\mathbf{H}_{\text{MISE},r} = \underset{\mathbf{H} \in \mathcal{H}}{\arg\min} \ \text{MISE}_r(\mathbf{H})$$

and

$$\mathbf{H}_{\text{AMISE},r} = \underset{\mathbf{H} \in \mathcal{H}}{\arg\min} \ \text{AMISE}_r(\mathbf{H}).$$

If we denote $D_\mathbf{H} = \frac{\partial}{\partial \text{vech}\mathbf{H}}$, then using matrix differential calculus yields (Magnus and Neudecker (1999))

$$D_\mathbf{H}\text{AMISE}_r(\mathbf{H}) = -(2n)^{-1} |\mathbf{H}|^{-1/2} \text{tr} \left\{ (\mathbf{H}^{-1})^r V(D^r K) \right\} \mathbf{D}_d^T \text{vec}\mathbf{H}^{-1}$$
$$+ n^{-1} |\mathbf{H}|^{-1/2} r \left(-\mathbf{D}_d^T \text{vec}(\mathbf{H}^{-1} V(D^r K)\mathbf{H}^{-1}) \right)$$
$$+ \frac{\beta_2^2}{2} \mathbf{\Psi}_{4+2r} \text{vech}\mathbf{H}.$$

Unfortunately, there is not any explicit solution for the equation

$$D_{\mathbf{H}}\text{AMISE}_r(\mathbf{H}) = \mathbf{0} \tag{7.6}$$

(with an exception of $d = 2$, $r = 0$ and a diagonal bandwidth matrix $\mathbf{H}$, see, *e.g.*, Wand and Jones (1995)).
Nevertheless the following lemma holds.

Lemma 7.1.

$$\text{AIV}_r(\mathbf{H}_{\text{AMISE},r}) = \frac{4}{d + 2r}\text{AISB}_r(\mathbf{H}_{\text{AMISE},r}). \tag{7.7}$$

Proof. See Complements for a proof.   □

It can be shown (Chacón and Duong (2010)) that

$$\mathbf{H}_{\text{AMISE},r} = \mathbf{C}_{0,r}n^{-2/(d+2r+4)} = O(n^{-2/(d+2r+4)}\mathbf{J}_d)$$

and

$$\text{AMISE}_r(\mathbf{H}_{\text{AMISE},r}) = O\left(n^{-4/(d+2r+4)}\right).$$

Since $\mathbf{H}_{\text{AMISE},r}$ and $\mathbf{H}_{\text{MISE},r}$, respectively, cannot be found in practice, the data-driven methods have been developed. Tarn Duong's PhD thesis (Duong (2004)) provides a comprehensive survey of bandwidth matrix selection methods. These methods are also widely discussed in the papers Chacón and Duong (2010); Duong (2004); Duong and Hazelton (2005b); Sain *et al.* (1994); Wand and Jones (1994) *etc.* Some papers (Horová *et al.* (2008b, 2012b); Vopatová *et al.* (2010); Horová and Vopatová (2011)) focus on constrained parameterization of the bandwidth matrix such as a diagonal matrix and the attention has been paid to visualization using bivariate functional surfaces – for more see Sec. 7.4.

The performance of bandwidth matrix selectors can be assessed by its relative rate of convergence. We generalize the definition for the relative rate of convergence for the univariate case to the multivariate one.

Definition 7.1. Let $\widehat{\mathbf{H}}_r$ be a data-driven bandwidth matrix selector. We say that $\widehat{\mathbf{H}}_r$ converges to $\mathbf{H}_{\text{AMISE},r}$ with a relative rate $n^{-\alpha}$ if

$$\text{vech}(\widehat{\mathbf{H}}_r - \mathbf{H}_{\text{AMISE},r}) = O(J_{d^*}n^{-\alpha})\text{vech}\mathbf{H}_{\text{AMISE},r}. \tag{7.8}$$

7.3.1 *Cross-validation method*

Cross-validation methods $CV_r(H)$ (Duong and Hazelton (2005b); Chacón and Duong (2012)) aim to estimate $MISE_r$. $CV_r(H)$ employs the objective function

$$CV_r(\mathbf{H}) = (-1)^r \mathrm{tr} \left\{ \frac{1}{n^2} \sum_{i,j=1}^{n} D^{2r}(K_{\mathbf{H}} * K_{\mathbf{H}})(\mathbf{X}_i - \mathbf{X}_j) \right.$$

$$\left. - \frac{2}{n(n-1)} \sum_{\substack{i,j=1 \\ i \neq j}}^{n} D^{2r} K_{\mathbf{H}}(\mathbf{X}_i - \mathbf{X}_j) \right\}, \quad (7.9)$$

$CV_r(\mathbf{H})$ is an unbiased estimate of $MISE_r(\mathbf{H}) - \mathrm{tr}V(D^r f)$.
Let

$$\widehat{\mathbf{H}}_{CV_r} = \underset{\mathbf{H} \in \mathcal{H}}{\arg\min} \ CV_r(\mathbf{H}).$$

It can be shown that relative rate of convergence to $\mathbf{H}_{MISE,r}$ is $n^{-d/(2d+4r+8)}$ (Duong and Hazelton (2005b); Chacón and Duong (2012)).

7.3.2 *Plug-in methods*

Plug-in methods were generalized to the multivariate case in Wand and Jones (1994). Duong (2004) developed full bandwidth matrix selectors for density estimate (*i.e.* $r = 0$) in the following way.
Let the plug-in criterion be

$$PI(\mathbf{H}) = n^{-1} V(K) |\mathbf{H}|^{-1/2} + \frac{\beta_2^2}{4} \mathrm{vech}^T \mathbf{H} \widehat{\mathbf{\Psi}}_4 \mathrm{vech} \mathbf{H}.$$

This is the $AMISE_0$ with $\mathbf{\Psi}_4$ replaced by $\widehat{\mathbf{\Psi}}_4$. Thus we aim to find $\widehat{\mathbf{H}}_{PI}$, the minimizer of PI

$$\widehat{\mathbf{H}}_{PI} = \underset{\mathbf{H} \in \mathcal{H}}{\arg\min} \ PI(\mathbf{H}).$$

In order to find this matrix we need to compute $\widehat{\mathbf{\Psi}}_4$. It could be done through kernel estimates of entries of the matrix $\mathbf{\Psi}_4$ (see Duong (2004)). The same idea can be used for the kernel gradient estimate where $\mathbf{\Psi}_{4+2r}$ should be estimated. The relative rate of convergence to $\mathbf{H}_{MISE,r}$ is $n^{-2/(d+2r+6)}$ when $d \geq 2$ (see, *e.g.*, Chacón and Duong (2010); Chacón and Duong (2012)).

7.3.3 Reference rule

Let f be a d-variate standard normal density $f \sim N(\boldsymbol{\mu}, \boldsymbol{\Sigma})$ and K be the d-dimensional Gaussian kernel $K(\mathbf{x}) = \phi_{\mathbf{I}}(\mathbf{x})$. Then in accordance with a univariate case

$$\widehat{\mathbf{H}}_{\text{REF}} = \left(\frac{4}{n(d+2)}\right)^{2/(d+4)} \widehat{\boldsymbol{\Sigma}}, \tag{7.10}$$

where $\widehat{\boldsymbol{\Sigma}}$ is the empirical estimate of a covariance matrix

$$\widehat{\boldsymbol{\Sigma}} = \frac{1}{n-1} \sum_{i=1}^{n} (\mathbf{X}_i - \overline{\mathbf{X}})(\mathbf{X}_i - \overline{\mathbf{X}})^T, \quad \text{with } \overline{\mathbf{X}} = \frac{1}{n} \sum_{i=1}^{n} \mathbf{X}_i.$$

7.3.4 Maximal smoothing principle

The idea of maximal smoothing for a univariate density can be easily extended to a multivariate case (see Terrell (1990)). An oversmoothed bandwidth matrix takes the form

$$\widehat{\mathbf{H}}_{\text{MS}} = \left(\frac{(d+8)^{(d+6)/2} \pi^{2/d} V(K)}{16n\Gamma\left(\frac{d+8}{2}\right)(d+2)}\right)^{2/(d+4)} \widehat{\boldsymbol{\Sigma}}, \tag{7.11}$$

where $\widehat{\boldsymbol{\Sigma}}$ is the empirical estimate of a sample covariance matrix and Γ is the gamma function.

7.3.5 Iterative method

Similarly, as in the univariate case we develop a method based on Lemma 7.1 in the sense that the solution of $D_{\mathbf{H}}\text{AMISE}_r(\mathbf{H}) = 0$ is equivalent to solving the equation (7.7). But $\text{AISB}_r(\mathbf{H})$ depends on the unknown density f. Thus we use a suitable estimate of $\text{AISB}_r(\mathbf{H})$. The equation (7.7) can be rewritten as

$$(d+2r)n^{-1}|\mathbf{H}|^{-1/2}\text{tr}\left\{(\mathbf{H}^{-1})^r V(D^r K)\right\}$$
$$- \beta_2^2 \text{vech}^T \mathbf{H} \boldsymbol{\Psi}_{4+2r} \text{vech}\mathbf{H} = 0. \tag{7.12}$$

Let us denote

$$\Lambda(\mathbf{z}) = (K * K * K * K - 2K * K * K + K * K)(\mathbf{z}),$$
$$\Lambda_{\mathbf{H}}(\mathbf{z}) = |\mathbf{H}|^{-1/2}\Lambda(\mathbf{H}^{-1/2}\mathbf{z}).$$

Then the unbiased estimate of $\text{AISB}_r(\mathbf{H})$ can be considered as

$$\overline{\text{AISB}}_r(\mathbf{H}) = \int \|(K_{\mathbf{H}} * \widehat{D^r f})(\mathbf{x}, \mathbf{H}) - \widehat{D^r f}(\mathbf{x}, \mathbf{H})\|^2 d\mathbf{x}.$$

Thus the estimate of $\mathrm{AISB}_r(\mathbf{H})$ can be proposed as

$$\widehat{\mathrm{AISB}}_r(\mathbf{H}) = \mathrm{tr}\left\{ \frac{(-1)^r}{n^2} \sum_{\substack{i,j=1 \\ i\neq j}}^{n} D^{2r}\Lambda_{\mathbf{H}}(\mathbf{X}_i - \mathbf{X}_j) \right\}.$$

Now, instead of the equation (7.12) we aim to solve the equation

$$(d+2r)n^{-1}|\mathbf{H}|^{-1/2}\mathrm{tr}\left\{ (\mathbf{H}^{-1})^r V(D^r K) \right\}$$
$$- 4\mathrm{tr}\left\{ \frac{(-1)^r}{n^2} \sum_{\substack{i,j=1 \\ i\neq j}}^{n} D^{2r}\Lambda_{\mathbf{H}}(\mathbf{X}_i - \mathbf{X}_j) \right\} = 0. \quad (7.13)$$

Remark 7.1. The bandwidth matrix selection method based on the equation (7.13) will be called *Iterative method* (IT method). The bandwidth estimate will be denoted $\widehat{\mathbf{H}}_{\mathrm{IT}_r}$. This approach was developed in the paper Horová *et al.* (2012a).

This method was originally proposed for the bandwidth matrix selection for a bivariate density provided that the bandwidth matrix is diagonal (Horová *et al.* (2008b, 2012b), see also Sec. 7.4). As concerns a kernel gradient estimator the above-mentioned method for a diagonal matrix was investigated in Vopatová *et al.* (2010) and for a full matrix in Horová and Vopatová (2011).

Remark 7.2. In the following we assume that K is a standard normal density $\phi_{\mathbf{I}}$. Thus $\Lambda(\mathbf{z}) = \phi_{4\mathbf{I}}(\mathbf{z}) - 2\phi_{3\mathbf{I}}(\mathbf{z}) + \phi_{2\mathbf{I}}(\mathbf{z})$ and $\beta_2 = 1$. We are going to discuss statistical properties of iterative method which show the rationality of it.

Let $\Gamma_r(\mathbf{H})$ stand for the left hand side of (7.12) and $\widehat{\Gamma}_r(\mathbf{H})$ for the left hand side of (7.13).

Theorem 7.2. *Let the assumptions $(A_1) - (A_3)$ be satisfied and $K = \phi_{\mathbf{I}}$. Then*

$$E(\widehat{\Gamma}_r(\mathbf{H})) = \Gamma_r(\mathbf{H}) + o(||\mathrm{vec}\mathbf{H}||^{5/2}),$$
$$\mathrm{var}(\widehat{\Gamma}_r(\mathbf{H})) = 32n^{-2}|\mathbf{H}|^{-1/2}||\mathrm{vec}\mathbf{H}||^{-2r}V(\mathrm{vec}D^{2r}\Lambda)V(f)$$
$$+ o(n^{-2}|\mathbf{H}|^{-1/2}||\mathrm{vec}\mathbf{H}||^{-2r}).$$

Proof. See Horová *et al.* (2012a). $\square$

As far as the convergence rate of IT method is concerned we inspired with AMSE lemma (Duong (2004); Duong and Hazelton (2005a)). The following theorem takes place.

Theorem 7.3. *Let the assumptions* (A_1) *–* (A_3) *be satisfied and* $K = \phi_{\mathbf{I}}$. *Then*

$$\mathrm{MSE}\{\mathrm{vech}\widehat{\mathbf{H}}_{\mathrm{IT}_r}\} = O\left(n^{-\min\{d+8,12\}/(d+2r+4)}\mathbf{J}_{d^*}\right) \times$$
$$\times \, \mathrm{vech}\mathbf{H}_{\mathrm{AMISE},r}\mathrm{vech}^T\mathbf{H}_{\mathrm{AMISE},r}.$$

Proof. Proof of theorem can be found in Horová *et al.* (2012a). $\qquad\square$

Corollary 7.1. *The convergence rate to* $\mathbf{H}_{\mathrm{AMISE},r}$ *is* $n^{-\min\{d,4\}/(2d+4r+8)}$ *for the IT method.*

Remark 7.3. *(Computational aspects)*
The equation (7.13) can be rewritten as

$$|\widehat{\mathbf{H}}_{\mathrm{IT}_r}|^{1/2}4\,\mathrm{tr}\left\{\frac{(-1)^r}{n}\sum_{\substack{i,j=1\\i\neq j}}^{n} D^{2r}\Lambda_{\widehat{\mathbf{H}}_{\mathrm{IT}_r}}(\mathbf{X}_i - \mathbf{X}_j)\right\}$$
$$= (d + 2r)\mathrm{tr}\left\{(\widehat{\mathbf{H}}_{\mathrm{IT}_r}^{-1})^r V\left(D^r K\right)\right\}.$$

This equation represents a nonlinear equation for d^* unknown entries of $\widehat{\mathbf{H}}_{\mathrm{IT}_r}$. In order to find all these entries we need additional $d^* - 1$ equations. Below, we describe a possibility of obtaining these equations. We adopt a similar idea as in the case of a diagonal matrix (see also Terrell (1990); Scott (1992); Duong *et al.* (2008); Horová and Vopatová (2011)). We explain this approach for the case $d = 2$ with a matrix

$$\widehat{\mathbf{H}}_{\mathrm{IT}_r} = \begin{pmatrix} \hat{h}_{11,r} & \hat{h}_{12,r} \\ \hat{h}_{12,r} & \hat{h}_{22,r} \end{pmatrix}.$$

Let $\widehat{\mathbf{\Sigma}}$ be a sample covariance matrix

$$\widehat{\mathbf{\Sigma}} = \begin{pmatrix} \hat{\sigma}_{11}^2 & \hat{\sigma}_{12} \\ \hat{\sigma}_{12} & \hat{\sigma}_{22}^2 \end{pmatrix}.$$

The initial estimates of entries of $\widehat{\mathbf{H}}_{\mathrm{IT}_r}$ can be chosen as

$$\hat{h}_{11,r} = \hat{h}_{1,r}^2 = (\hat{\sigma}_{11}^2)^{(12+r)/12}\, n^{(r-4)/12},$$
$$\hat{h}_{22,r} = \hat{h}_{2,r}^2 = (\hat{\sigma}_{22}^2)^{(12+r)/12}\, n^{(r-4)/12},$$
$$\hat{h}_{12,r} = \mathrm{sign}\,\hat{\sigma}_{12}|\hat{\sigma}_{12}|^{(12+r)/12}\, n^{(r-4)/12}.$$

For details see Horová and Vopatová (2011).
Hence

$$\hat{h}_{22,r} = \left(\frac{\hat{\sigma}_{22}^2}{\hat{\sigma}_{11}^2}\right)^{(12+r)/12} \hat{h}_{11,r}, \tag{7.14}$$

$$\hat{h}_{12,r}^2 = \left(\frac{\hat{\sigma}_{12}^2}{\hat{\sigma}_{11}^2}\right)^{(12+r)/12} \hat{h}_{11,r} \tag{7.15}$$

and further

$$|\widehat{\mathbf{H}}_{\mathrm{IT}_r}| = \hat{h}_{11,r}^2 \left((\hat{\sigma}_{11}\hat{\sigma}_{22})^{(12+r)/6} - \hat{\sigma}_{12}^{(12+r)/6}\right) \Big/ \hat{\sigma}_{11}^{(12+r)/3}$$

$$= \hat{h}_{11,r}^2 S(\hat{\sigma}_{ij}).$$

Thus we arrive at the equation for the unknown $\hat{h}_{11,r}$

$$4\hat{h}_{11,r} \sqrt{S(\hat{\sigma}_{ij})} \mathrm{tr}\left\{\frac{(-1)^r}{n} \sum_{\substack{i,j=1 \\ i\neq j}}^{n} D^{2r}\Lambda_{\widehat{\mathbf{H}}_{\mathrm{IT}_r}}(\mathbf{X}_i - \mathbf{X}_j)\right\} \tag{7.16}$$

$$= (d+2r)\mathrm{tr}\left\{(\widehat{\mathbf{H}}_{\mathrm{IT}_r}^{-1})^r V (D^r K)\right\}.$$

This approach is very important for computational aspects of solving equation (7.13). Putting the equations (7.14), (7.15) to (7.16) forms one nonlinear equation for the unknown $\hat{h}_{11}$ and it can be solved by means of an appropriate iterative numerical method. This procedure gives the name of the proposed method. Evidently, this approach is computationally much faster than a general minimization process.

7.4   A special case for bandwidth selection

In this section we focus on a problem of a data-driven choice of a bandwidth matrix for bivariate kernel density estimate. Bivariate kernel density estimation provided that the bandwidth matrix is diagonal is an excellent setting for understanding aspects of a multivariate kernel smoothing. Moreover, in this case we are able to clarify the process of the bandwidth matrix choice by using bivariate functional surfaces.

Thus let us suppose that $d = 2$, K be a product kernel and $\mathbf{H} \in \mathbf{H}_{\mathcal{D}}$. Then $\mathrm{AMISE}_0(\mathbf{H})$ takes the form

$$\mathrm{AMISE}_0(\mathbf{H}) = \frac{V(K)}{h_1 h_2} + \frac{\beta_2^2}{4}(h_1^4 \psi_{4,0} + 2h_1^2 h_2^2 \psi_{2,2} + h_2^4 \psi_{0,4}),$$

where

$$\psi_{k,l} = \int \int \left(\frac{\partial^2 f}{\partial x_1^2}\right)^{k/2} \left(\frac{\partial^2 f}{\partial x_2^2}\right)^{l/2} dx_1 dx_2, \ k, l = 0, 2, 4, \ k + l = 4.$$

In such a case the equation $D_{\mathbf{H}}\text{AMISE}_0 = 0$ has a unique solution (Wand and Jones (1995))

$$\begin{aligned}
h_{1,\text{AMISE},0}^2 &= \left[\frac{\psi_{0,4}^{3/4} V(K)}{n\beta_2 \psi_{4,0}^{3/4}\left(\psi_{2,2} + \psi_{0,4}^{1/2}\psi_{4,0}^{1/2}\right)}\right]^{1/3} \\
h_{2,\text{AMISE},0}^2 &= \left[\frac{\psi_{4,0}^{3/4} V(K)}{n\beta_2 \psi_{0,4}^{3/4}\left(\psi_{2,2} + \psi_{0,4}^{1/2}\psi_{4,0}^{1/2}\right)}\right]^{1/3}.
\end{aligned} \tag{7.17}$$

Now, the equation (7.7) can be written as

$$\text{AIV}(\mathbf{H}_{\text{AMISE}}) = 2\text{AISB}(\mathbf{H}_{\text{AMISE}}). \tag{7.18}$$

For the sake of simplicity the notation $\mathbf{H}_{\text{AMISE}}$ instead of $\mathbf{H}_{\text{AMISE},0}$ will be used in this paragraph. The suitable approximation of (7.18) is

$$\frac{V(K)}{nh_1 h_2} = \frac{2}{n^2 h_1 h_2} \sum_{\substack{i,j=1 \\ i \neq j}}^{n} \Lambda\left(\frac{X_{i_1} - X_{j_1}}{h_1}, \frac{X_{i_2} - X_{j_2}}{h_2}\right), \tag{7.19}$$

where

$$\Lambda(z_1, z_2) = (K * K * K * K - 2K * K * K + K * K)(z_1, z_2).$$

It is well-known that a visualization is an important component of a non-parametric data analysis and thus we describe the proposed method by means of bivariate functionals.

Figure 7.4 shows the shape of the functional $\widehat{\Gamma}_0(h_1, h_2) = nV(K) - 2\sum_{\substack{i,j=1 \\ i \neq j}}^{n} \Lambda\left(\frac{X_{i_1} - X_{j_1}}{h_1}, \frac{X_{i_2} - X_{j_2}}{h_2}\right)$ and the point we are seeking for lies on a curve $\widehat{\Gamma}_0(h_1, h_2) = 0$, i.e., it is an intersection of the surface $\widehat{\Gamma}_0(h_1, h_2)$ and the co-ordinate plane $z = 0$. Obviously, there is not a unique solution, thus we need another relationship between h_1 and h_2 to find the unique solution. We proposed $M1$ and $M2$ methods described below.

7.4.1 *M1 method*

This method uses Scott's rule (Scott (1992))

$$\hat{h}_i = \hat{\sigma}_i n^{-1/6}, \ i = 1, 2$$

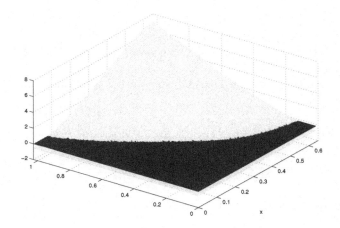

Fig. 7.4   3-D plot of $\Gamma_0(h_1, h_2)$.

which yields the second equation of the form

$$h_2 = \hat{c}h_1, \ \hat{c} = \frac{\hat{\sigma}_2}{\hat{\sigma}_1},$$

and $\hat{\sigma}$ can be estimated by, $e.g.$, a sample deviation.   Thus we arrive
at the system

$$M1 \begin{cases} \widehat{\Gamma}_0(h_1, h_2) = 0 \\ h_1 = \hat{c}h_2. \end{cases} \tag{7.20}$$

Figure 7.5 demonstrates the solution of the system $M1$.

7.4.2   M2 method

This method can be considered as a hybrid of the biased cross-validation
(Duong and Hazelton (2005b); Sain et $al.$ (1994)) and the plug-in method
(Wand and Jones (1994)).

It is easy to show that

$$h_{2,\text{AMISE}}^4 \psi_{0,4} = h_{1,\text{AMISE}}^4 \psi_{4,0}.$$

Assuming that K is a product kernel we can express the estimates of $\psi_{0,4}$
and $\psi_{4,0}$ as

$$\widehat{\psi}_{0,4} = \frac{1}{n^2} \sum_{i,j=1}^{n} \left(\frac{\partial^2 K_{\mathbf{H}}}{\partial x_2^2} * \frac{\partial^2 K_{\mathbf{H}}}{\partial x_2^2} \right) (\mathbf{X}_i - \mathbf{X}_j),$$

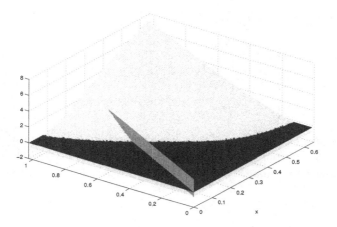

Fig. 7.5   Graphical representation of $M1$.

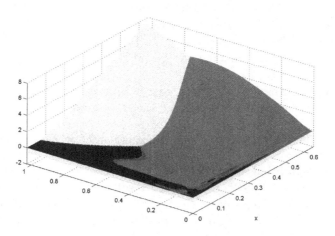

Fig. 7.6   Graphical representation of $M2$.

$$\widehat{\psi}_{4,0} = \frac{1}{n^2} \sum_{i,j=1}^{n} \left(\frac{\partial^2 K_{\mathbf{H}}}{\partial x_1^2} * \frac{\partial^2 K_{\mathbf{H}}}{\partial x_1^2} \right) (\mathbf{X}_i - \mathbf{X}_j).$$

The second method is described by the system

$$
M2 \quad \begin{cases} \widehat{\Gamma}_0(h_1, h_2) = 0 \\[2mm] h_2^4 \sum_{i,j=1}^{n} \left(\frac{\partial^2 K_{\mathbf{H}}}{\partial x_2^2} * \frac{\partial^2 K_{\mathbf{H}}}{\partial x_2^2} \right) (\mathbf{X}_i - \mathbf{X}_j) \\[4mm] \quad = h_1^4 \sum_{i,j=1}^{n} \left(\frac{\partial^2 K_{\mathbf{H}}}{\partial x_1^2} * \frac{\partial^2 K_{\mathbf{H}}}{\partial x_1^2} \right) (\mathbf{X}_i - \mathbf{X}_j). \end{cases} \tag{7.21}
$$

This system can be solved by Newton's method or by secant method. In Fig. 7.6 the graphs of the surfaces given by $M2$ are presented. The solution $(\hat{h}_{1,\mathrm{AMISE}}, \hat{h}_{2,\mathrm{AMISE}})$ is an intersection of both these surfaces and plane $z = 0$.

7.5 Simulations

To test the effectiveness of the proposed estimator, we simulated its performance against the least squares cross-validation method. The simulation is based on 100 replications of 6 mixture normal bivariate densities, labelled A to F. Means and covariance matrices of these distributions were generated randomly. The formulas are given in Table 7.1. Densities A and B are unimodal, C and D are bimodal and E and F are trimodal. Their contour plots are in Fig. 7.7.

The sample size of $n = 100$ was used in all replications. For each estimated density derivative we calculated the Integrated Square Error (ISE)

$$
\mathrm{ISE}_1\{\widehat{Df}(\cdot, \mathbf{H})\} = \int \|\widehat{Df}(\mathbf{x}, \mathbf{H}) - Df(\mathbf{x})\|^2 d\mathbf{x}
$$

over all 100 replications and displayed the logarithm of results in a boxplot in Fig. 7.8. Here "ITER" denotes the results for the iterative method, "LSCV" stands for the results of the Least Square Cross-Validation method (7.9) and "MISE" is a tag for the results obtained by minimizing (7.5). The variance of each estimator can be accurately gauged by the whiskers of the plot.

7.6 Application to real data

In the use of smoothing methods in data analysis, an important question is which observed features – such as a local extreme – are really there. Chaudhuri and Marron (1999) introduced the SiZer (Significant Zero) method for

Table 7.1 Normal mixture densities

Density	Formula $N(\text{vec}^T\boldsymbol{\mu}, \text{vech}^T\boldsymbol{\Sigma})$
A	$N\big((-0.2686, -1.7905), (7.9294, -10.0673, 22.1150)\big)$
B	$N\big((-0.6847, 2.6963), (16.9022, 9.8173, 6.0090)\big)$
C	$\frac{1}{2}N\big((0.3151, -1.6877), (0.1783, -0.1821, 1.0116)\big)$ $+\frac{1}{2}N\big((1.1768, 0.3731), (0.2414, -0.8834, 4.2934)\big)$
D	$\frac{1}{2}N\big((1.8569, 0.1897), (1.5023, -0.9259, 0.8553)\big)$ $+\frac{1}{2}N\big((0.3349, -0.2397), (2.3050, 0.8895, 1.2977)\big)$
E	$\frac{1}{3}N\big((0.0564, -0.9041), (0.9648, -0.8582, 0.9332)\big)$ $+\frac{1}{3}N\big((-0.7769, 1.6001), (2.8197, -1.4269, 0.9398)\big)$ $+\frac{1}{3}N\big((1.0132, 0.4508), (3.9982, -3.7291, 5.5409)\big)$
F	$\frac{1}{3}N\big((2.2337, -2.9718), (0.6336, -0.9279, 3.1289)\big)$ $+\frac{1}{3}N\big((-4.3854, 0.5678), (2.1399, -0.6208, 0.7967)\big)$ $+\frac{1}{3}N\big((1.5513, 2.2186), (1.1207, 0.8044, 1.0428)\big)$

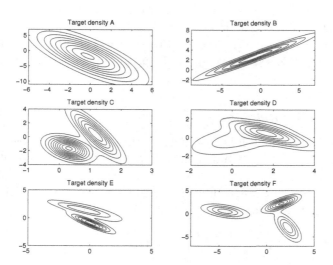

Fig. 7.7 Contour plots for target densities.

finding structure in smooth data. Duong *et al.* (2008) proposed a frame-
work for feature significance in d-dimensional data which combines kernel
density derivative estimators and hypothesis tests for modal regions. For
the gradient and curvature estimators distributional properties are given,

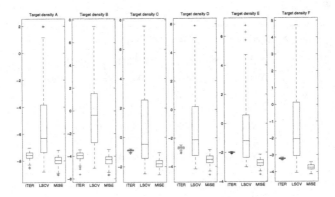

Fig. 7.8   Box plots for log(ISE).

and pointwise tests extend the two-dimensional feature significance ideas of Godtliebsen *et al.* (2002).

For this reason we also developed the method for kernel gradient estimation and in this section we apply the proposed method also to four real data sets.

7.6.1   *Old Faithful data*

We started with the well-known "Old Faithful" data set Simonoff (1996), which contains characteristics of 222 eruptions of the 'Old Faithful Geyser' in Yellowstone National Park, USA, during August 1978 and August 1979. Kernel density and the first derivative estimates using the Gaussian kernel based on following bandwidth matrices

$$\widehat{\mathbf{H}}_{\mathrm{IT}_0} = \begin{pmatrix} 0.0703 & 0.7281 \\ 0.7281 & 9.801 \end{pmatrix}, \qquad \widehat{\mathbf{H}}_{\mathrm{IT}_1} = \begin{pmatrix} 0.2388 & 3.006 \\ 3.006 & 50.24 \end{pmatrix}.$$

are displayed in Fig. 7.9. The curves $\frac{\partial f}{\partial x_1} = 0$, $\frac{\partial f}{\partial x_2} = 0$ and their intersections confirm the existence of local extremes.

7.6.2   *UNICEF data*

The second data set is taken from UNICEF – The State of the Worlds Children 2003. It contains 72 pairs of observation for countries with GNI less than 1 000 US dollars per capita in 2001. X_1 variable describes the under-

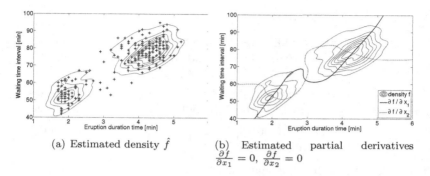

(a) Estimated density $\hat{f}$

(b) Estimated partial derivatives $\frac{\partial f}{\partial x_1} = 0$, $\frac{\partial f}{\partial x_2} = 0$

Fig. 7.9 'Old Faithful' data contour plots $- \hat{f}$ and $\widehat{Df}$.

five mortality rate, *i.e.*, the probability of dying between birth and exactly five years of age expressed per 1 000 live births, and X_2 is a life expectancy at birth, *i.e.*, the number of years newborn children would live if subject to the mortality risks prevailing for the cross-section of population at the time of their birth UNICEF (2003). These data have also been analyzed in Duong and Hazelton (2005b).

Bandwidth matrices for the estimated density $\hat{f}$ and its gradient $\widehat{Df}$ are following

$$\widehat{\mathbf{H}}_{\mathrm{IT}_0} = \begin{pmatrix} 1112.0 & -138.3 \\ -138.3 & 24.20 \end{pmatrix}, \qquad \widehat{\mathbf{H}}_{\mathrm{IT}_1} = \begin{pmatrix} 2426 & -253.7 \\ -253.7 & 38.38 \end{pmatrix}.$$

Figure 7.10 illustrates the use of the iterative bandwidth matrices for the 'Children mortality' data set.

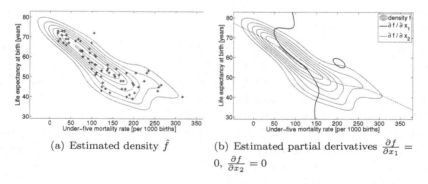

(a) Estimated density $\hat{f}$

(b) Estimated partial derivatives $\frac{\partial f}{\partial x_1} = 0$, $\frac{\partial f}{\partial x_2} = 0$

Fig. 7.10   'Children mortality' data contour plots $- \hat{f}$ and $\widehat{Df}$.

7.6.3 *Swiss bank notes*

We also analyze Swiss bank notes data set from Simonoff (1996). It contains measurements of the bottom margin and diagonal length of the 100 real Swiss bank notes and 100 forged Swiss bank notes.

Figure 7.11 presents kernel estimates of the joint distribution of the bottom margin and diagonal length of the bills using the bandwidth matrices

$$\widehat{\mathbf{H}}_{\mathrm{IT}_0} = \begin{pmatrix} 0.1227 & -0.0610 \\ -0.0610 & 0.0781 \end{pmatrix}, \qquad \widehat{\mathbf{H}}_{\mathrm{IT}_1} = \begin{pmatrix} 0.6740 & -0.3159 \\ -0.3159 & 0.4129 \end{pmatrix}.$$

The bills with longer diagonal and shorter bottom margin correspond to real bills.

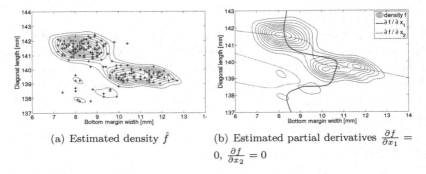

(a) Estimated density $\hat{f}$

(b) Estimated partial derivatives $\frac{\partial f}{\partial x_1} = 0$, $\frac{\partial f}{\partial x_2} = 0$

Fig. 7.11   Swiss bank notes data contour plots – $\hat{f}$ and $\widehat{Df}$.

7.6.4 *Iris flower*

As the last example, we estimated the three-dimensional density of the Iris flower data set (Anderson (1935)). The data set consists of 50 samples from each of three species of Iris flowers (Iris setosa, Iris virginica and Iris versicolor). Four features were measured from each sample originally, they are the length and the width of sepal and petal.

This data set was introduced in Fisher (1936) as an example of discriminant analysis. Based on the combination of the four features, Fisher developed a linear discriminant model to determine which species they are. We used following three variables to estimate their density:

x_1 – sepal width [cm],
x_2 – petal length [cm],

x_3 – petal width [cm].

For an estimation of the optimal bandwidth matrix we used two methods; the reference rule and the iterative method. We obtained following bandwidth matrices:

$$\widehat{\mathbf{H}}_{\mathrm{REF}} = \begin{pmatrix} 0.0426 & -0.0739 & -0.0273 \\ -0.0739 & 0.6986 & 0.2904 \\ -0.0273 & 0.2904 & 0.1302 \end{pmatrix},$$

$$\widehat{\mathbf{H}}_{\mathrm{IT}_0} = \begin{pmatrix} 0.0288 & -0.0493 & -0.0181 \\ -0.0493 & 0.4775 & 0.1989 \\ -0.0181 & 0.1989 & 0.0893 \end{pmatrix}.$$

Figure 7.12 presents contour plots of the kernel estimates of the joint density using the bandwidth matrices $\widehat{\mathbf{H}}_{\mathrm{REF}}$ and $\widehat{\mathbf{H}}_{\mathrm{IT}_0}$.

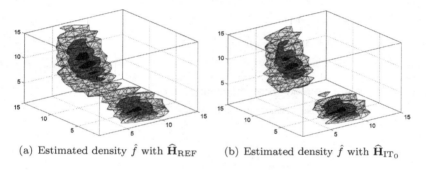

(a) Estimated density $\hat{f}$ with $\widehat{\mathbf{H}}_{\mathrm{REF}}$ (b) Estimated density $\hat{f}$ with $\widehat{\mathbf{H}}_{\mathrm{IT}_0}$

Fig. 7.12 Iris flower data contour plots of $\hat{f}$.

The using of $\widehat{\mathbf{H}}_{\mathrm{IT}_0}$ in the smoothing process seems to be more appropriate for the discriminant analysis. For more about this example see Vopatová (2012).

7.7   Use of MATLAB toolbox

The toolbox can be downloaded from the web page
`http://www.math.muni.cz/english/science-and-research/`
`developed-software/232-matlab-toolbox.html`.

7.7.1   *Running the program*

The *Start menu* (Fig. 7.13) for kernel estimation of two-dimensional density
is called up by the command `ksbivardens`.

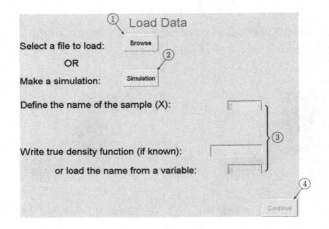

Fig. 7.13   Start menu.

You can skip this menu by typing input data as an argument
`ksbivardens(X)`, where the matrix X should have the size $2 \times n$, where
n is the sample size. If we know also the original density f (for example for simulated data), we can set it as the next argument. For more
see `help ksbivardens`. After the execution of this command, the window
in Fig. 7.16 is called up directly.

In the *Start menu*, you have several possibilities how to define input
data. You can load it from a file (button ①) or simulate data (button
②). In the fields ③ you can list your variables in the current workspace
to define input data. If your workspace is empty, these fields are nonactive.
If you know the true density of the sample, you can write it to the text field
or load it from a variable. If you need to simulate a sample, press button

$\binom{2}{}$. Then the menu for simulation (Fig. 7.14) is called up. This application generates a random sample from a two-dimensional normal mixture density.

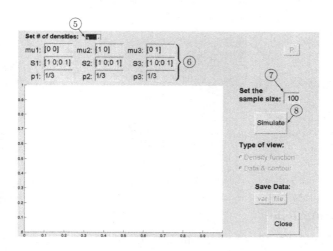

Fig. 7.14   Simulation menu.

In the *Simulation menu*, first set the number of components of the mixture by $\binom{5}{}$. You can choose from 1 to 5 components of the normal mixture density. In the fields $\binom{6}{}$ specify the parameters (mean, variance matrix and proportion) of each component. By $\binom{7}{}$ specify the sample size. You can simulate a sample by pressing button $\binom{8}{}$ and then see the result (Fig. 7.15).

By clicking on $\binom{9}{}$ you can print the current plot to a new figure. In the fields $\widehat{10}$ you switch the type of view between the data plot with contours and the 3-D plot of the density function. You can also save the obtained data to variables and then as a file by using buttons $\widehat{11}$. If you have done the simulation, press button $\widehat{12}$. The *Simulation menu* will be closed and you will be returned to the *Start menu*. In this menu, you can redefine the input data. If you want to continue, press button $\binom{4}{}$. The menu will be closed and the *Basic menu* (see the next subsection) will be called up.

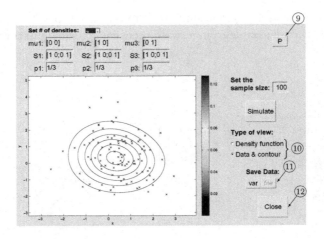

Fig. 7.15   Simulation menu – results.

7.7.2   *Main figure*

This menu (Fig. 7.16) was called up from the *Start menu* or directly from the command line (see `help ksbivardens`). If the original density is known, in ⑭ you can switch the type of view between the data plot with contours and the 3-D plot of the density function. By clicking on ⑮

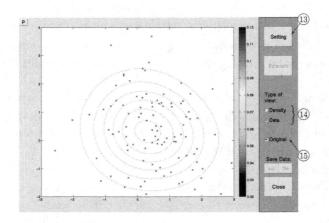

Fig. 7.16   Basic menu.

you can show or hide contours of original density. If the original density is unknown, only data are plotted. Use button (13) to continue.

7.7.3 Setting the parameters

Button (13) calls up the menu for setting the parameters (Fig. 7.17) which will be used for bivariate kernel density estimation.

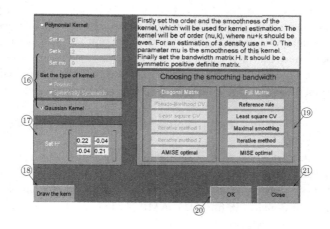

Fig. 7.17   Setting of parameters.

In the array (16), you can set a type of the kernel. Implicit setting is the Gaussian kernel or it can be changed to polynomial kernel. In this case, set the type of kernel (product or spherically symmetric) and set the order of the kernel (ν, k), where $\nu + k$ should be even, for density estimation $\nu = 0$ is used. The parameter μ is the smoothness of this kernel. If you want to draw the kernel, use button (18). In the array (17), specify the bandwidth matrix **H**. The matrix should be symmetric and positive definite. If these conditions are not satisfied, the application writes an error message. The implicit setting is

$$\widehat{\mathbf{H}} = \widehat{\mathbf{\Sigma}} n^{-1/3}, \tag{7.22}$$

where $\widehat{\mathbf{\Sigma}}$ is the empirical estimate of the covariance matrix and n is the number of observations. This formula (known as "Reference rule") is based on the assumption of normal density and Gaussian kernel (see §7.3.3).

The terms of the bandwidth matrix can be set manually or you can use one of the buttons in the array ⑲. There are two groups of methods:

(1) **Diagonal matrix** – in this case, it is supposed that the bandwidth matrix is diagonal. Because of computational aspects, the used methods are developed for product polynomial kernels. For other types of kernel the buttons are not active. There are five buttons:

- *Pseudo-likelihood CV* – represents the pseudo-likelihood cross-validation method described in Horová *et al.* (2009). This bivariate method is a straightforward extension of univariate pseudo-likelihood CV (see Cao *et al.* (1994)).
- *Least squares CV* – the method is based on the paper Sain *et al.* (1994), where some research on LSCV selector for diagonal matrix was carried out.
- *Iterative method 1* – it represents the proposed $M1$ method described in §7.4.1.
- *Iterative method 2* – this means the second proposed iterative method $M2$, which is explained in §7.4.2.
- *AMISE optimal* – this button is active only in the case of known original density, it finds the AMISE optimal diagonal bandwidth matrix by the formula (7.17).

(2) **Full matrix** – in this case, the full bandwidth matrix and Gaussian kernel $\phi_{\mathbf{I}}$ are supposed. There are also five buttons:

- *Reference rule* – represents the implicit setting based on the multiplication of the covariance matrix estimate, see the formula (7.22).
- *Least squares CV* – it minimizes the CV error function (7.9) for $r = 0$.
- *Maximal smoothing* – represents the maximal smoothing principle described in §7.3.4.
- *Iterative method* – it represents the proposed iterative method for full matrix described in §7.3.5.
- *MISE optimal* – this button is active only in the case of known mixture of normal distributions as an original density. It estimates the MISE optimal bandwidth matrix by minimizing (7.5).

To confirm the setting use ⑳, to close the window, use ㉑.

7.7.4 *The final estimation*

If you set all the values in the window for setting parameters (see Fig. 7.17) and if you want to go to the final estimation of the density, confirm your setting by button ⟨20⟩. It calls up the *Basic menu*, where all buttons are active already.

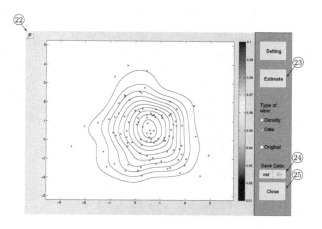

Fig. 7.18   Final kernel density estimate.

By clicking on button ⟨23⟩ the relevant kernel density estimate is drawn. You can again add contours for the known original density by ⟨15⟩ and switch between types of view in ⟨14⟩. By clicking on ⟨22⟩ you can print the current plot to a new figure. You can also save data to variables and then as a file by using buttons ⟨24⟩. Button ⟨25⟩ ends all applications.

7.8   Complements

Sketch of the proof of Lemma 7.1:

Proof.   Consider the equation (7.6) and multiply it from the left by $\frac{1}{2}\text{vech}^T H$.

Then

$$(4n)^{-1}|\mathbf{H}|^{-1/2}\text{vech}^T H \text{tr}\left\{(\mathbf{H}^{-1})^r V(D^r K)\right\} D_d^T \text{vec}\mathbf{H}^{-1}$$
$$+(2n)^{-1}|\mathbf{H}|^{-1/2}r\text{vech}^T H \left(D_d^T \text{vec}(\mathbf{H}^{-1}V(D^r K))\mathbf{H}^{-1}\right)$$
$$= \frac{\beta_2^2}{4}\text{vech}^T H \mathbf{\Psi}_{4+2r}\text{vech}\mathbf{H}.$$

The right hand side of this equation is $AISB_r$. Further if we use the facts on matrix calculus we arrive at the formula (7.7). $\qquad\square$

Notation

Kernels

Kernel estimates at the given point

Errors

Bandwidths and their estimates

Estimates of a density

Reliability assessment

Survival analysis

Kernel regression

Other

For any function or quantity Q symbol $\widehat{Q}$ denotes its estimate.

Let a_n and b_n be sequences of real numbers
$a_n = o(b_n)$ if and only if $\lim\limits_{n \to \infty} \left| \frac{a_n}{b_n} \right| = 0$.
$a_n = O(b_n)$ if and only if $\limsup\limits_{n \to \infty} \left| \frac{a_n}{b_n} \right| < \infty$.

The integral $\int$ stands for $\int\limits_{-\infty}^{\infty}$ or $\int\limits_{\mathbb{R}^d}$ unless is stated otherwise.

Bibliography

Akaike, H. (1970). Statistical predictor information, *Annals of the Institute of Statistical Mathematics* **22**, pp. 203–217.

Akaike, H. (1974). Markovian representation of stochastic-processes and its application to analysis of autoregressive moving average processes, *Annals of The Institute of Statistical Mathematics* **26**, 3, pp. 363–387.

Altman, N. and Léger, C. (1995). Bandwidth selection for kernel distribution function estimation, *Journal of Statistical Planning and Inference* **46**, pp. 195–214.

Anderson, E. (1935). The irises of the Gaspe Peninsula, *Bulletin of the American Iris Society* **59**, pp. 2–5.

Anderson, R. (2007). *The credit scoring toolkit: theory and practice for retail credit risk management and decision automation* (Oxford University Press).

Azzalini, A. (1981). A note on the estimation of a distribution function and quantiles by a kernel method, *Biometrika* **68**, 1, pp. 326–328.

Bowman, A. (1984). An alternative method of cross-validation for the smoothing of density estimates, *Biometrika* **71**, 2, pp. 353–360.

Bowman, A., Hall, P. and Prvan, T. (1998). Bandwidth selection for the smoothing of distribution functions, *Biometrika* **85**, 4, pp. 799–808.

Bowman, A. W. and Azzalini, A. (1997). *Applied smoothing techniques for data analysis: The kernel approach with S-plus illustrations*, Oxford statistical science series (Oxford University Press, USA).

Brillinger, D. (2001). *Time series: data analysis and theory*, Classics in applied mathematics (Society for Industrial and Applied Mathematics).

Cao, R., Cuevas, A. and González Manteiga, W. (1994). A comparative study of several smoothing methods in density estimation, *Computational Statistics and Data Analysis* **17**, 2, pp. 153–176.

Chacón, J. E. and Duong, T. (2010). Multivariate plug-in bandwidth selection with unconstrained pilot bandwidth matrices, *Test* **19**, 2, pp. 375–398.

Chacón, J. E. and Duong, T. (2012). Bandwidth selection for multivariate density derivative estimation, with applications to clustering and bump hunting, *ArXiv e-prints* URL http://adsabs.harvard.edu/abs/2012arXiv1204. 6160C.

Chacón, J. E., Duong, T. and Wand, M. P. (2011). Asymptotics for general multivariate kernel density derivative estimators, *Statistica Sinica* **21**, 2, pp. 807–840.

Chaudhuri, P. and Marron, J. S. (1999). SiZer for exploration of structures in curves, *Journal of the American Statistical Association* **94**, 447, pp. 807–823.

Cheng, M.-Y. (1997). Boundary aware estimators of integrated density derivative products, *Journal of the Royal Statistical Society. Series B (Methodological)* **59**, 1, pp. 191–203.

Chiu, S. (1990). Why bandwidth selectors tend to choose smaller bandwidths, and a remedy, *Biometrika* **77**, 1, pp. 222–226.

Chiu, S. (1991). Some stabilized bandwidth selectors for nonparametric regression, *Annals of Statistics* **19**, 3, pp. 1528–1546.

Chu, C. K. and Marron, J. S. (1991). Choosing a kernel regression estimator, *Statistical Science* **6**, 4, pp. 404–419.

Cleveland, W. S. (1979). Robust locally weighted regression and smoothing scatterplots, *Journal of the American Statistical Association* **74**, 368, pp. 829–836.

Cline, D. and Hart, J. (1991). Kernel estimation of densities of discontinuous derivatives, *Statistics* **22**, pp. 69–84.

Collet, D. (1997). *Modelling survival data in medical research* (Chapman & Hall/CRC, Boca Raton, London, New York, Washington D.C.).

Coppock, D. S. (2002). Why lift? www.dmreview.com/news/5329-1.html.

Craven, P. and Wahba, G. (1979). Smoothing noisy data with spline functions - estimating the correct degree of smoothing by the method of generalized cross-validation, *Numerische Mathematik* **31**, 4, pp. 377–403.

Deheuvels, P. (1977). Estimation nonparametrique de la densité par histogrammes generalisés, *Rev. Statist. Appl.* **35**, pp. 5–42.

DeLong, E., DeLong, D. and D.L. Clarke-Pearson, D. L. (1988). Comparing the areas under two or more correlated receiver operating characteristic curves: a nonparametric approach, *Biometrics* **44**, 3, pp. 837–845.

Devroye, L. and Lugosi, G. (1997). Nonasymptotic universal smoothing factors, kernel complexity and Yatracos classes, *The Annals of Statistics* **25**, 6, pp. 2626–2637.

Doleželová, H., Šlampa, P., Ondrová, B., Gombošová, A., Sovadinová, Š., Novotný, T., Bolčák, K., Růžičková, J., Hynková, L. and Forbelská, M. (2008). The impact of PET with 18FDG in radiotherapy treatment planning and in the prediction in patients with cervix carcinoma - results of pilot study, *Neoplasma* **55**, 5, pp. 437–441.

Doubek, M., Jungová, A., Brejcha, M., Panovská, A., Brychtová, Y., Pospíšil, Z. and Mayer, J. (2009). Alemtuzumab in chronic lymphocytic leukemia treatment: retrospective analysis of outcome according to cytogenetics, *Vnitřní lékařství (Internal Medicine, in Czech)* **55**, 6, pp. 549–554.

Droge, B. (1996). Some comments on cross-validation, Tech. Rep. 1994-7, Humboldt Universitaet Berlin, URL http://ideas.repec.org/p/wop/humbsf/1994-7.html.

Duong, T. (2004). *Bandwidth selectors for multivariate kernel density estimation*, Ph.D. thesis, School of Mathematics and Statistics, University of Western Australia.

Duong, T. (2007). ks: Kernel density estimation and kernel discriminant analysis for multivariate data in R, *Journal of Statistical Software* **21**, 7, pp. 1–16.

Duong, T., Cowling, A., Koch, I. and Wand, M. P. (2008). Feature significance for multivariate kernel density estimation, *Computational Statistics & Data Analysis* **52**, 9, pp. 4225–4242.

Duong, T. and Hazelton, M. (2005a). Convergence rates for unconstrained bandwidth matrix selectors in multivariate kernel density estimation, *Journal of Multivariate Analysis* **93**, 2, pp. 417–433.

Duong, T. and Hazelton, M. (2005b). Cross-validation bandwidth matrices for multivariate kernel density estimation, *Scandinavian Journal of Statistics* **32**, 3, pp. 485–506.

Engel, E. (1857). Die Productions und Consumtionsverhältnisse des Königreichs Sachsen, *Zeitschrift des statistischen Bureaus des Königlich Sächsischen Ministerium des Inneren* **8,9**.

Epanechnikov, V. A. (1969). Non-parametric estimation of a multivariate probability density, *Theory of Probability and its Applications* **14**, 1, pp. 153–158.

Eubank, R. (1988). *Spline smoothing and nonparametric regression*, 1st edn. (Dekker, New York).

Fan, J. (1992). Design-adaptive nonparametric regression, *Journal of the American Statistical Association* **87**, 420, pp. 998–1004.

Fan, J. and Gijbels, I. (1995). Data-driven bandwidth selection in local polynomial fitting: Variable bandwidth and spatial adaptation, *Journal of the Royal Statistical Society. Series B (Methodological)* **57**, 2, pp. pp. 371–394.

Fang, K., Kotz, S. and Ng, K. (1990). *Symmetric multivariate and related distributions* (Chapman and Hall, London; New York).

Fisher, R. A. (1922). On the mathematical foundations of theoretical statistics, *Philosophical Transactions of the Royal Society of London, A* **222**, pp. 309–368.

Fisher, R. A. (1932). *Statistical methods for research workers* (Oliver & Boyd, Edinburgh).

Fisher, R. A. (1936). The use of multiple measurements in taxonomic problems, *Annals of Human Genetics* **7**, 2, pp. 179–188.

Forbelská, M. (2007). Elliptically contoured models in ROC analysis, in *S.Co.2007, Fifth Conference, Book of Short Papers*, pp. 243–248.

Gasser, T. and Müller, H.-G. (1979). Kernel estimation of regression functions, in T. Gasser and M. Rosenblatt (eds.), *Smoothing Techniques for Curve Estimation, Lecture Notes in Mathematics*, Vol. 757 (Springer Berlin / Heidelberg), pp. 23–68.

Gasser, T., Müller, H.-G. and Mammitzsch, V. (1985). Kernels for nonparametric curve estimation, *Journal of the Royal Statistical Society. Series B (Methodological)* **47**, 2, pp. 238–252.

Godtliebsen, F., Marron, J. S. and Chaudhuri, P. (2002). Significance in scale space for bivariate density estimation, *Journal of Computational and*

Graphical Statistics **11**, 1, pp. 1–21.

González-Manteiga, W., Cao, R. and Marron, J. S. (1996). Bootstrap selection of the smoothing parameter in nonparametric hazard rate estimation, *Journal of the American Statistical Association* **91**, 435, pp. 1130–1140.

Granovsky, B. and Müller, H.-G. (1989). On the optimality of a class of polynomial kernel functions, *Journal of Statistics and Decisions* **7**, pp. 301–312.

Granovsky, B. and Müller, H.-G. (1991). Optimizing kernel methods - a unifying variational principle, *International Statistical Review* **59**, 3, pp. 373–388.

Granovsky, B. L., Müller, H.-G. and Pfeifer, C. (1995). Some remarks on optimal kernel functions, *Journal of Statistics and Decisions* **13**, 2, pp. 101–116.

Graunt, J. (1662). *Natural and political observations made upon the bills of mortality* (Martyn London).

Hall, P. and Marron, J. S. (1987). On the amount of noise inherent in bandwidth selection for a kernel density estimator, *The Annals of Statistics* **15**, 1, pp. 163–181.

Hall, P. and Park, B. U. (2002). New methods for bias correction at endpoints and boundaries, *The Annals of Statistics* **30**, 5, pp. 1460–1479.

Hall, P. and Wehrly, T. E. (1991). A geometrical method for removing edge effects from kernel-type nonparametric regression estimators, *Journal of the American Statistical Association* **86**, 415, pp. pp. 665–672.

Hand, D. and Henley, W. (1997). Statistical classification methods in consumer credit scoring: a review, *Journal of the Royal Statistical Society, Series A.* **160**, 3, pp. 523–541.

Härdle, W. (1990). *Applied Nonparametric Regression*, 1st edn. (Cambridge University Press, Cambridge).

Härdle, W., Hall, P. and Marron, J. (1988). How far are automatically chosen regression smoothing parameters from their optimum, *Journal of the American Statistical Association* **83**, 401, pp. 86–95.

Härdle, W., Marron, J. S. and Wand, M. P. (1990). Bandwidth choice for density derivatives, *Journal of the Royal Statistical Society. Series B (Methodological)* **52**, 1, pp. 223–232.

Härdle, W., Müller, M., Sperlich, S. and Wewatz, A. (2004). *Nonparametric and Semiparametric Models*, 1st edn. (Heidelberg: Springer).

Hjort, N. and Jones, M. (1996). Locally parametric nonparametric density estimation, *The Annals of Statistics* **24**, 4, pp. 1619–1647.

Horová, I. (1996). Gegenbauer polynomials, optimal kernels and Stancu operators, in *Approximation Theory and Function Series* (Budapest: János Bolyai Mathematical Society), pp. 227–235.

Horová, I. (1997). Boundary kernels, in *Summer schools MATLAB 94, 95* (Brno: Masaryk University), pp. 17–24.

Horová, I. (2000). Some remarks on kernels, *Journal of Computational Analysis and Applications* **2**, pp. 253–263.

Horová, I. (2002). *Optimization problems connected with kernel estimates*, Electrical and Computer Engineering Series (WSEAS Press), pp. 334–339.

Horová, I., Koláček, J. and Vopatová, K. (2009). Bandwidth matrix choice for bivariate kernel density estimates, in *Book of short papers* (MU Brno), pp.

22–25.

Horová, I., Koláček, J. and Vopatová, K. (2012a). Full bandwidth matrix selectors for gradient kernel density estimate, *Computational Statistics & Data Analysis*, submitted.

Horová, I., Koláček, J. and Vopatová, K. (2012b). Visualization and bandwidth matrix choice, *Communications in Statistics – Theory and Methods* **41**, 4, pp. 759–777.

Horová, I., Koláček, J., Zelinka, J. and El-Shaarawi, A. H. (2008a). Smooth estimates of distribution functions with application in environmental studies, in *Advanced topics on mathematical biology and ecology* (WSEAS Press), pp. 122–127.

Horová, I., Koláček, J., Zelinka, J. and Vopatová, K. (2008b). Bandwidth choice for kernel density estimates, in *Proceedings IASC, Yokohama: IASC, 2008*, pp. 542–551.

Horová, I., Pospíšil, Z. and Zelinka, J. (2007). Semiparametric estimation of hazard function for cancer patients, *Sankhya: The Indian Journal of Statistics* **69**, 3, pp. 494–513.

Horová, I., Pospíšil, Z. and Zelinka, J. (2009). Hazard function for cancer patients and cancer cell dynamics, *Journal of Theoretical Biology* **258**, 3, pp. 437–443.

Horová, I., Vieu, P. and Zelinka, J. (2002). Optimal choice of nonparametric estimates of a density and of its derivatives, *Statistics & Decisions* **20**, 4, pp. 355–378.

Horová, I. and Vopatová, K. (2011). Kernel gradient estimate, in F. Ferraty (ed.), *Recent Advances in Functional Data Analysis and Related Topics* (Springer-Verlag Berlin Heidelberg), pp. 177–182.

Horová, I. and Zelinka, J. (2007a). Contribution to the bandwidth choice for kernel density estimates, *Computational Statistics* **22**, 1, pp. 31–47.

Horová, I. and Zelinka, J. (2007b). Kernel estimation of hazard functions for biomedical data sets, in W. Härdle, Y. Mori and P. Vieu (eds.), *Statistical Methods for Biostatistics and Related Fields*, Mathematics and Statistics (Springer-Verlag Berlin Heidelberg), pp. 64–86.

Horová, I., Zelinka, J. and Budíková, M. (2006). Kernel estimates of hazard functions for carcinoma data sets, *Environmetrics* **17**, 3, pp. 239–255.

Hougaard, P. (2001). *Analysis of Multivariate Survival Data* (Springer-Verlag, New York, Berlin, Heidelberg).

Hurt, J. (1992). Statistical methods for survival data analysis, in *Proceedings ROBUST'92, Eds. J.Antoch & G.Dohnal*, pp. 54–74.

Jiang, J. and Marron, J. S. (2003). Sizer for censored density and hazard estimation, Preprint.

Jones, M. and Sheather, S. (1991). Using nonstochastic terms to advantage in kernel-based estimation of integrated squared density derivatives, *Statistics & Probability Letters* **11**, 6, pp. 511–514.

Jones, M. C. (1993). Simple boundary correction for kernel density estimation, *Statistics and Computing* **3**, pp. 135–146.

Jones, M. C. and Kappenman, R. F. (1991). On a class of kernel density estimate bandwidth selectors, *Scandinavian Journal of Statistics* **19**, 4, pp. 337–349.

Jones, M. C., Marron, J. S. and Park, B. U. (1991). A simple root n bandwidth selector, *The Annals of Statistics* **19**, 4, pp. 1919–1932.

Kaplan, E. I. and Meier, P. V. (1958). Nonparametric estimation from incomplete observations, *Journal of the American Statistical Association* **53**, 282, pp. 457–481.

Karunamuni, R. and Alberts, T. (2005a). On boundary correction in kernel density estimation, *Statistical Methodology* **2**, pp. 191–212.

Karunamuni, R. and Alberts, T. (2005b). A generalized reflection method of boundary correction in kernel density estimation, *Canad. J. Statist.* **33**, pp. 497–509.

Karunamuni, R. and Alberts, T. (2006). A locally adaptive transformation method of boundary correction in kernel density estimation, *J. Statist. Planning and Inference* **136**, pp. 2936–2960.

Karunamuni, R. and Zhang, S. (2008). Some improvements on a boundary corrected kernel density estimator, *Statistics & Probability Letters* **78**, pp. 497–507.

Kim, C., Bae, W., Choi, H. and Park, B. U. (2005). Non-parametric hazard function estimation using the Kaplan–Meier estimator, *Journal of Nonparametric Statistics* **17**, 8, pp. 937–948.

Klein, J. and Moeschberger, M. (2003). *Survival analysis: techniques for censored and truncated data*, Statistics for biology and health (Springer).

Koláček, J. (2002). Kernel estimation of the regression function - bandwidth selection, in *Summer School DATASTAT01 Proceedings FOLIA* (MU Brno), pp. 129–138.

Koláček, J. (2005). *Kernel Estimation of the Regression Function (in Czech)*, Ph.D. thesis, Masaryk University, Brno.

Koláček, J. (2008). An improved estimator for removing boundary bias in kernel cumulative distribution function estimation (in Czech), in *Proceedings in Computational Statistics COMPSTAT'08*, pp. 549–556.

Koláček, J. (2008). Plug-in method for nonparametric regression, *Computational Statistics* **23**, 1, pp. 63–78.

Koláček, J. and Karunamuni, R. J. (2009). On boundary correction in kernel estimation of ROC curves, *Austrian Journal of Statistics* **38**, 1, pp. 17–32.

Koláček, J. and Karunamuni, R. J. (2011). A generalized reflection method for kernel distribution and hazard functions estimation, *Journal of Applied Probability and Statistics* **6**, 2, pp. 73–85.

Koláček, J. and Poměnková, J. (2006). A comparative study of boundary effects for kernel smoothing, *Austrian Journal of Statistics* **35**, 2, pp. 281–289.

Koláček, J. and Řezáč, M. (2010). Assessment of scoring models using information value, in *19th International Conference on Computational Statistics, Paris France, August 22-27, 2010 Keynote, Invited and Contributed Papers*, pp. 1191–1198.

Koláček, J. and Řezáč, M. (2011). Quality measures for predictive scoring models, in *Proceedings ASMDA 2011* (Roma, Italy), pp. 720–727.

Koláček, J. and Zelinka, J. (2012). MATLAB toolbox, URL http://www.math. muni.cz/english/science-and-research/developed-software/ 232-matlab-toolbox.html.

Koziol, J. A. and Green, S. B. (1976). Cramer–von Mises statistic for randomly censored data, *Biometrika* **63**, 3, pp. 465–474.

Kozusko, F. and Bajzer, Z. (2003). Combining gompertzian growth and cell population dynamics, *Mathematical Biosciences* **185**, 2, pp. 153–167.

Krzanowski, W. J. and Hand, D. J. (2009). *ROC curves for continuous data*, Monographs on statistics and applied probability (CRC Press).

Kullback, S. and Leibler, R. A. (1951). On information and sufficiency, *The Annals of Mathematical Statistics* **22**, 1, pp. 79–86.

Lejeune, M. (1985). Estimation non-paramétrique par noyaux: régression polynomiale mobile, *Revue de Statistique Applique* **33**, 3, pp. 43–67.

Lejeune, M. and Sarda, P. (1992). Smooth estimators of distribution and density functions, *Computational Statistics & Data Analysis* **14**, pp. 457–471.

Li, K.-C. (1985). From Stein's unbiased risk estimates to the method of generalized cross validation, *The Annals of Statistics* **13**, 4, pp. 1352–1377.

Lloyd, C. (1998). Using smoothed receiver operating characteristic curves to summarize and compare diagnostic systems, *Journal of the American Statistical Association* **93**, 444.

Lloyd, C. (2002). Estimation of a convex ROC curve, *Statistics & Probability Letters* **59**, 1, pp. 99–111.

Lloyd, C. and Yong, Z. (1999). Kernel estimators of the ROC curve are better than empirical, *Statistics & Probability Letters* **44**, pp. 221–228.

Loader, C. R. (1996). Local likelihood density estimation, *The Annals of Statistics* **24**, 4, pp. 1602–1618.

Lusted, L. B. (1971). Signal detectability and medical decision-making, *Science* **171**, 3977, pp. 1217–1219, doi:10.1126/science.171.3977.1217, URL http: //www.sciencemag.org/content/171/3977/1217.short.

Mack, Y. P. and Müller, H.-G. (1988). Convolution type estimators for nonparametric regression, *Statistics & Probability Letters* **7**, 3, pp. 229–239.

Magnus, J. R. and Neudecker, H. (1999). *Matrix differential calculus with applications in statistics and econometrics*, 2nd edn. (John Wiley & Sons).

Mallows, C. (1973). Some comments on C_p, *Technometrics* **15**, 4, pp. 661–675.

Mammitzsch, V. (1985). The fluctuation of kernel estimators under certain moment conditions, in *Proc. ISI (ISI'85)*, pp. 17–18.

Marron, J. S. (1996). A personal view of smoothing and statistics, in W. Härdle and M. Schimek (eds.), *Statistical Theory and Computational Aspects of Smoothing, Contributions to Statistics* (Physica-Verlag), pp. 1–9.

Marron, J. S. and Nolan, D. (1988). Canonical kernels for density-estimation, *Statistics & Probability Letters* **7**, 3, pp. 195–199.

Marron, J. S. and Padgett, W. J. (1987). Asymptotically optimal bandwidth selection for kernel density estimators from randomly right-censored samples, *Annals of Statistics* **15**, 4, pp. 1520–1535.

Marron, J. S. and Ruppert, D. (1994). Transformations to reduce boundary bias in kernel density estimation, *Journal of the Royal Statistical Society. Series*

B (Methodological) **56**, 4, pp. 653–671.

Mielniczuk, J. (1986). Some asymptotic properties of kernel estimators of a density function in case of censored data, *Annals of Statistics* **14**, 2, pp. 766–773.

Minnotte, M. C. (1993). *A test of mode existence with application to multimodality*, Ph.D. thesis, Rice University, TX, USA.

Müller, H., Stadtmüller, U. and Schmitt, T. (1987). Bandwidth choice and confidence-intervals for derivatives of noisy data, *Biometrika* **74**, 4, pp. 743–749.

Müller, H.-G. (1987). Weighted local regression and kernel methods for nonparametric curve fitting, *Journal of the American Statistical Association* **82**, 397, pp. 231–238.

Müller, H.-G. (1988). *Nonparametric regression analysis of longitudinal data* (Springer, New York).

Müller, H.-G. (1991). Smooth optimum kernel estimators near endpoints, *Biometrika* **78**, 3, pp. 521–530.

Müller, H. G. and Wang, J. L. (1990a). Locally adaptive hazard smoothing, *Probability Theory and Related Fields* **85**, 4, pp. 523–538.

Müller, H. G. and Wang, J. L. (1990b). Nonparametric analysis of changes in hazard rates for censored survival data – an alternative to change-point models, *Biometrika* **77**, 2, pp. 305–314.

Müller, H. G. and Wang, J. L. (1994). Hazard rate estimation under random censoring with varying kernels and bandwidths, *Biometrics* **50**, 1, pp. 61–76.

Nadaraya, E. A. (1964). On estimating regression, *Theory of Probability and its Applications* **9**, 1, pp. 141–142.

Nelson, W. (1972). Theory and applications of hazard plotting for censored data, *Technometrics* **14**, pp. 945–966.

Nielsen, J. P. and Linton, O. B. (1995). Kernel estimation in a nonparametric marker dependent hazard model, *Annals of Statistics* **23**, 5, pp. 1735–1748.

Nováková, J. (2009). *Kernel cumulative distribution function estimates (in Czech)*, Master's thesis, Masaryk University, Brno.

Parzen, E. (1962). On estimation of a probability density function and mode, *The Annals of Mathematical Statistics* **33**, 3, pp. 1065–1076.

Patil, P. N. (1993a). Bandwidth choice for nonparametric hazard rate estimation, *Journal of Statistical Planning and Inference* **35**, 1, pp. 15–30.

Patil, P. N. (1993b). On the least-squares cross-validation bandwidth in hazard rate estimation, *Annals of Statistics* **21**, 4, pp. 1792–1810.

Patil, P. N., Wells, M. T. and Marron, J. S. (1994). Some heuristics of kernel based estimators of ratio functions, *Nonparametrics Statistics* **4**, pp. 283–289.

Pepe, M. (2004). *The statistical evaluation of medical tests for classification and prediction*, Oxford statistical science series (Oxford University Press).

Priestley, M. B. and Chao, M. T. (1972). Non-parametric function fitting, *Journal of the Royal Statistical Society. Series B (Methodological)* **34**, 3, pp. 385–392.

Ramlau-Hansen, C. H. (1983). Counting processes intensities by means of kernel functions, *The Annals of Statistics* **11**, 2, pp. 453–466.

Reiss, R.-D. (1981). Nonparametric estimation of smooth distribution functions, *Scand. J. Stat., Theory Appl.* **8**, pp. 116–119.

Ren, H., X.H.Zhou and Liang, H. (2004). A flexible method for estimating the ROC curve, *Journal of Applied Statistics* **31**, 7, pp. 773–784.

Řezáč, F. and Řezáč, M. (2011). How to measure the quality of credit scoring models, *Czech Journal of Economics and Finance* **61**, 5, pp. 486–507.

Řezáč, M. (2007). *Kernel density estimates (in Czech)*, Ph.D. thesis, Masaryk University, Brno.

Řezáč, M. and Koláček, J. (2010). On aspects of quality indexes for scoring models, in *19th International Conference on Computational Statistics, Paris France, August 22-27, 2010 Keynote, Invited and Contributed Papers*, pp. 1517–1524.

Řezáč, M. and Koláček, J. (2011). Adjusted empirical estimate of information value for credit scoring models, in *Proceedings ASMDA 2011* (Roma, Italy), pp. 1162–1169.

Rice, J. (1984). Bandwidth choice for nonparametric regression, *Annals of Statistics* **12**, 4, pp. 1215–1230.

Rosenblatt, M. (1956). Remarks on some nonparametric estimates of a density function, *The Annals of Mathematical Statistics* **27**, 3, pp. 832–837.

Sain, S., Baggerly, K. and Scott, D. (1994). Cross-validation of multivariate densities, *Journal of the American Statistical Association* **89**, 427, pp. 807–817.

Sarda, P. (1993). Smoothing parameter selection for smooth distribution functions, *Journal of Statistical Planning and Inference* **35**, pp. 65–75.

Sarda, P. and Vieu, P. (1991). Smoothing parameter selection in hazard estimation, *Statistics & Probability Letters* **11**, 5, pp. 429–434.

Scheffé, H. (1959). *The analysis of variance*, Wiley Classics Library (Wiley-Interscience Publication).

Schuster, E. (1985). Incorporating support constraints into nonparametric estimators of densities, *Communications in Statistics-Theory end Methods* **14**, 5, pp. 1123–1136.

Scott, D. W. (1992). *Multivariate density estimation: theory, practice, and visualization* (Wiley).

Scott, D. W. and Terrell, G. R. (1987). Biased and unbiased cross-validation in density estimation, *Journal of the American Statistical Association* **82**, 400, pp. 1131–1146.

Searle, S. (1987). *Linear models for unbalanced data* (Wiley).

Seber, G. (1977). *Linear regression analysis* (Wiley).

Seifert, B. and Gasser, T. (1996). Variance properties of local polynomials and ensuing modifications, in W. Härdle and M. Schimek (eds.), *Statistical Theory and Computational Aspects of Smoothing, Contributions to Statistics* (Physica-Verlag), pp. 50–79.

Shibata, R. (1981). An optimal selection of regression variables, *Biometrika* **68**, 1, pp. 45–54.

Siddiqi, N. (2006). *Credit risk scorecards: developing and implementing intelligent credit scoring*, Wiley and SAS Business Series (Wiley).

Silverman, B. W. (1985). Some aspects of the spline smoothing approach to non-parametric regression curve fitting, *Journal of the Royal Statistical Society. Series B (Methodological)* **47**, pp. 1–52.

Silverman, B. W. (1986). *Density estimation for statistics and data analysis* (Chapman and Hall, London).

Simonoff, J. S. (1996). *Smoothing Methods in Statistics* (Springer-Verlag, New York).

Soumarová, R., Horová, H., Růžičková, J., Čoupek, P., Šlampa, P., Šeneklová, Z., Petráková, K., Budíková, M. and Horová, I. (2002). Local and distant failure in patients with stage i and ii carcinoma of the breast treated with breat-conserving surgery and radiation therapy (in Czech, English summary), *Radiační onkologie* **2**, 1, pp. 17–24.

Stoer, J. and Bulirsch, R. (2002). *Introduction to numerical analysis*, 3rd edn., Texts in applied mathematics (Springer).

Stone, C. J. (1977). Consistent nonparametric regression, *The Annals of Statistics* **5**, 4, pp. 595–620.

Stone, M. (1974). Cross-validatory choice and assessment of statistical predictions, *Journal of the Royal Statistical Society Series B-Statistical Methodology* **36**, 2, pp. 111–147.

Sturges, H. A. (1926). The choice of a class interval, *Journal of the American Statistical Association* **21**, 153, pp. pp. 65–66.

Szegő, G. (1939). *Orthogonal polynomials* (American Mathematical Society, Providence).

Tanner, M. A. and Wong, W. H. (1983). The estimation of the hazard function from randomly censored data by the kernel method, *Annals of Statistics* **11**, 3, pp. 989–993.

Tanner, M. A. and Wong, W. H. (1984). Data-based nonparametric estimation of the hazard function with applications to model diagnostics and exploratory analysis, *Journal of the American Statistical Association* **79**, 385, pp. 174–182.

Terrell, G. R. (1990). The maximal smoothing principle in density estimation, *Journal of the American Statistical Association* **85**, 410, pp. 470–477.

Terrell, G. R. and Scott, D. W. (1985). Oversmoothed nonparametric density estimates, *Journal of the American Statistical Association* **80**, 389, pp. 209–214.

Thernau, T. M. and Grambsch, P. M. (2001). *Modelling survival data. extending the Cox model* (Springer-Verlag, New York, Berlin, Heidelberg).

Thomas, L. (2009). *Consumer credit models: pricing, profit, and portfolios* (Oxford University Press).

Tukey, J. W. (1961). Curves as parameters, and touch estimation, in *Proceedings of the 4th Symposium on Mathematics, Statistics and Probability*, pp. 681–694.

Tukey, J. W. (1977). *Exploratory data analysis* (Addison-Wesley).

UNICEF (2003). The state of the worlds children 2003, http://www.unicef.org/sowc03/index.html.

Uzunogullari, U. and Wang, J. L. (1992). A comparison of hazard rate estimators for left truncated and right censored data, *Biometrika* **79**, 2, pp. 297–310.

Vopatová, K. (2012). K-web: KDE, http://k101.unob.cz/~vopatova/kde.htm.

Vopatová, K., Horová, I. and Koláček, J. (2010). Bandwidth choice for kernel density derivative, in *Proceedings of the 25th International Workshop on Statistical Modelling* (Glasgow, Scotland), pp. 561–564.

Wan, S. and Zhang, B. (2008). Comparing correlated ROC curves for continuous diagnostic tests under density ratio models, *Computational Statistics & Data Analysis* **53**, 1, pp. 233 – 245.

Wand, M. and Jones, M. (1993). Comparison of smoothing parameterizations in bivariate kernel density-estimation, *Journal of The American Statistical Association* **88**, 422, pp. 520–528.

Wand, M. and Jones, M. (1995). *Kernel smoothing* (Chapman and Hall, London).

Wand, M. P. and Jones, M. C. (1994). Multivariate plug-in bandwidth selection, *Computational Statistics* **9**, 2, pp. 97–116.

Watson, G. S. (1964). Smooth regression analysis, *Sankhya: The Indian Journal of Statistics, Series A* **26**, 4, pp. 359–372.

Watson, G. S. and Leadbetter, M. R. (1964). Hazard analysis I, *Biometrika* **51**, 1/2, pp. 175–184.

Whittle, P. (1958). On the smoothing of probability density functions, *Journal of the Royal Statistical Society. Series B* **55**, pp. 549–557.

Xu, K. (2003). How has the literature on Gini's index evolved in the past 80 years? *SSRN eLibrary* .

Yandell, B. (1983). Nonparametric inference for rates with censored survival data, *Annals of Statistics* **11**, 4, pp. 1119–1135.

Youndjé, E., Sarda, P. and Vieu, P. (1996). Optimal smooth hazard estimates, *Test* **5**, 2, pp. 379–394.

Zelinka, J. and Horová, I. (2001). Kernel estimates of a derivative of a regression function (in Czech), in *Proceedings ROBUST 2000, Eds. J.Antoch*, pp. 382–391.

Zhang, S. and Karunamuni, R. (1998). On kernel density estimation near endpoints, *Journal of Statistical Planning and Inference* **70**, pp. 301–316.

Zhang, S. and Karunamuni, R. (2000). On nonparametric density estimation at the boundary, *Journal Of Nonparametric Statistics* **12**, 2, pp. 197–221.

Zhang, S., Karunamuni, R. J. and Jones, M. C. (1999). An improved estimator of the density function at the boundary, *Journal of the American Statistical Association* **94**, 448, pp. 1231–1241.

Zhou, X. and Harezlak, J. (2002). Comparison of bandwidth selection methods for kernel smoothing of ROC curves, *Statistics in Medicine* **21**, 14, pp. 2045–2055.

Zhou, X., Obuchowski, N. and McClish, D. (2002). *Statistical methods in diagnostic medicine*, 1 (Wiley-Interscience).

Zou, K., Hall, W. and Shapiro, D. (1997). Smooth non-parametric receiver operating characteristic (ROC) curves for continuous diagnostic tests, *Statistics in Medicine* **16**, 19, pp. 2143–2156.

Index